IMAGES
of America

KNOXVILLE'S WNOX

On the Cover: This is a star-studded photograph of the "Class of 1947" WNOX *Mid-Day Merry-Go-Round.* See pages 18 and 19 for more information. (Courtesy of Patsy Robinson.)

IMAGES
of America

KNOXVILLE'S WNOX

Ed Hooper

ARCADIA
PUBLISHING

ISBN 978-0-7385-6653-5

Published by Arcadia Publishing
Charleston, South Carolina

Printed in the United States of America

Library of Congress Control Number: 2009931124

For all general information contact Arcadia Publishing at:
Telephone 843-853-2070
Fax 843-853-0044
E-mail sales@arcadiapublishing.com
For customer service and orders:
Toll-Free 1-888-313-2665

Visit us on the Internet at www.arcadiapublishing.com

This book is dedicated to Fred Shepherd, Lowell Blanchard, John Pirkle, "Your" Dave Young, and all the former employees of WNOX.

Contents

ACKNOWLEDGMENTS

I would like to thank those who contributed photographs to Knoxville's WNOX. This book would not be possible without the invaluable assistance of unofficial Knoxville radio archivist and historian "Your" Dave Young. Unless otherwise noted, all photographs were provided by him. Other contributors include John Pirkle, Bill Hayes, Connie Pritchett and the Pritchett family, John Hitch, Jan McCroskey, the Library of Congress, the Fred Shepherd family, Kendra Davis, Patsy Robertson, Phil Williams, the Cochran family, the E. W. Scripps Company, Edith Kirkland, the University of Tennessee Athletic Department, the Ale House, www.oidar.com by John Long, Ron Anderson, "Sunshine Slim" Sweet, and Roy Garrett of Roy's Record Shop.

Introduction

Knoxville broadcasting is home to legendary stations but none more than WNOX. In November 1921, when 16-year-old Knoxville native Stuart Adcock first signed the signal on the air as an employee of the Peoples Telephone and Telegraph, WNAV-AM became the eighth station in the United States to come into existence. Radio was still a grand experiment then. There were a few stations here and there across the nation, business models for the new medium were developing, and few really knew what potential lay in the electronic signal for music, information, and community service. In 1924, the young station was burned to the ground in a fire but returned a year later again under the direction of Peoples Telephone and Telegraph, when the call letters were changed from WNAV to WNOX.

Recognizing the value of the station to the community, Adcock applied to the U.S. Department of Commerce to get the letter "K" as the lead call letter in the station identification to spell "KNOX" but was refused, as Department of Commerce (DOC) regulations at the time restricted call letters beginning with "K" for stations west of the Mississippi and "W" for stations east of the Mississippi—unless licenses were already granted with the designation. WNOX was in a unique position of being a pioneer in broadcasting and mass communication for not only Knoxville, but the entire East Tennessee region. In fact, the airwaves at the time were so uncluttered the signal could be heard as far away as New York City.

Adcock sold WNOX in 1928 for $3,000 to the Sterchi Brothers, who leased it to Virgil Evans, and he moved it from the St. James Hotel on Wall Avenue to the Andrew Johnson Hotel on Gay Street. As radios became more popular and cheaper to own, the station saw a boom in listenership. It remained under the Sterchi Brothers' purview until 1932, when it was sold to the Liberty Life Insurance Company for $50,000. It operated out of the downtown St. James Hotel. Two years later, for $125,000, it became the second station in the nation to be purchased by Cincinnati-based Scripps Howard Radio, Inc.; they pushed the station onto the national stage.

The WNOX call letters were more than appropriate for Scripps Howard, Inc., who also owned the *Knoxville News Sentinel*—one of two leading newspapers in the city at the time. Radio in the beginning was a live production from the moment the microphone was turned on and quickly became a tourist attraction for people who wanted to see for themselves the "magic" of broadcasting. The station's power was increased to 10,000 watts, the signal moved to 990 AM, and the station affiliated with the national Columbia Broadcasting System (CBS) network.

Dick Westergaard and company moved the station from the Andrew Johnson Hotel to an abandoned tabernacle at 110 South Gay Street, which came equipped with its own auditorium able to accommodate the curious crowds. With a region like East Tennessee, which was so untouched by American popular culture and so rich in a unique musical heritage, Westergaard cultivated the idea of giving a platform to local musicians and hired Illinois broadcaster Lowell Blanchard in 1935 to drive into the hills and find talent for the station to audition. Blanchard became a fixture. Neither had the slightest idea the station's little programs would unknowingly give birth to an entire genre of American music and launch the careers of legendary entertainers.

The footprint of WNOX went far beyond its legendary musical programs. It was Knoxville's ear on the world in news and events from the first day it signed on the air. The station carried *Westinghouse's World Events* as an evening national news program. When Charles Lindberg's 18-month-old baby was kidnapped in March 1932, local and national interest was such that the station ran 24 hours a day for five days straight to keep people up to date on events surrounding it. When Scripps Howard took over the station, WNOX began utilizing United Press International's wire service. It was the first wire service to allow radio usage and allowed the station to implement a "floating news" policy, which meant it would interrupt programming if necessary to announce breaking news and devoted much to public affairs in East Tennessee.

In 1954, WNOX bought the old Whittle Springs Resort property and the hotel in North Knoxville. They built an auditorium, as they had also applied for a television license for WBIR. When they lost their bid for the television station, they converted it into a radio studio instead, and the building soon became a fixture in the city. The station featured great talent but also served as a training ground for many broadcasters, including this author, and the call letters were a coveted mark on radio résumés.

Because of its incredible story, WNOX's history has often been cherry-picked to feature only the station's most popular elements. This is the story of the station itself.

Overcoming that cherry-picked history would not have been possible without the invaluable assistance of radio historian and announcer "Your" Dave Young. The Scott County native and Knoxville broadcaster is an alumnus of both the WNOX and WIVK radio powerhouses. He entered the business as the pioneering generation was exiting and handing it to the succeeding one. Young learned from the elders and remembered the names and faces and their accomplishments. When times changed, he saw value in old station photographs of announcers and staff being discarded. Many were candid shots that showed what life was like in the world of Knoxville broadcasting. He kept notes on all of it—who was who, who started what—and collected discarded recordings and clippings from magazines and newspapers. Young kept them preserved and remembered to record his own generation's history and their accomplishments. The rest of the photographs in this book were gleaned from family members of musicians, former employees, and fans.

Writing about a pioneer organization like WNOX, I discovered there are two types of radio historians. One follows the signal on the dial, which moved frequently in the early days of radio, and the other follows the call letters by which it's known to people. This history is the latter and illustrates a station that not only appealed to Knoxville and East Tennessee but left behind an unrivaled national legacy that made it the unofficial headquarters and training ground for country music's greatest entertainers.

One

Country Music's Songwriting Legend

Arthur Q. Smith was born James Arthur Pritchett on December 9, 1909, in Griffin, Georgia. He hitchhiked his way into Knoxville in 1936 with a guitar and pocketful of songs he had written. He did odd jobs and worked as a bellhop at the Andrew Johnson Hotel in downtown Knoxville, hoping for a chance to audition for WNOX's *Mid-Day Merry-Go-Round*, which was a live variety program that actually featured all kinds of popular music of the day—from big band to gospel to a new "hillbilly" sound that was getting its first launch over the airwaves in America.

The latter music category soared in popularity with listeners, and the station could not play enough of it. James Pritchett got his audition and was hired as a rhythm guitarist for the house band within a month of moving to the city. He created the stage name Arthur Smith using his stepfather's last name and the Southern tradition of going by the middle name.

Old folk song standards, which were the popular tunes, soon wore out with listeners, however, and new songs were needed for performers to gain access to the stage and microphones of WNOX. Arthur Smith earned a little recognition for some of the songs he sang on WNOX's *Mid-Day Merry-Go-Round* and the once weekly *Tennessee Barn Dance* that featured nothing but the new hillbilly sound. When another songwriter named Arthur Smith at WBT radio in Charlotte, North Carolina, had a hit song with "Guitar Boogie," the station—in true broadcast style—held a contest to give Arthur Smith a middle name to distinguish him from the other. A local woman simply wrote the letter "Q" as an entry and won. Arthur Q. Smith was born.

The station gave Smith a 15-minute live music show known as *The Tennessee Cornshucker* program. He successfully used it as a songwriting launching pad and was prolific in his work. The songs were a new kind of sound built for radio, and the lyrics spoke to the heart of its listeners. After a stint as a soldier in World War II, he returned to Knoxville and picked up where he left off writing music.

Those were the Wild West days of country music. Songwriters were often looked upon as second-class citizens in the business and used mercilessly. Smith would sell his songs for anywhere from $5 to $25 each. It was a good sum of money, but Arthur was supporting a family and a serious drinking problem. He would sell most songs to performers who would put their names on the tunes as if they wrote them. Many became country music anthems that made fortunes for the entertainers in royalties. Some even took advantage of Smith's weakness and would put him up in a hotel and buy him liquor while he turned out song after song.

Among the hundreds of songs were "Wedding Bells Are Ringing In The Chapel," "I Overlooked An Orchid While Searching For A Rose," "Rainbow At Midnight," "I Wouldn't Change You If I Could," "Crying My Heart Out Over You," and "I Can't Stop Loving You." The last song was rewritten from a song Ruby Moody wrote titled "I Won't Stop Loving You." There were hundreds of others he was rumored to have written that helped form the baseline of a new country music sound, which became the standard of American popular music. He would walk the streets of Knoxville, duck into a bank, and write a song on a deposit slip or scratch it on a napkin or piece of trash he found in the road.

Those who bought and made the songs hits read like a who's who of entertainers: Bill Monroe, the Carter Family, Claude Boone, Ernest Tubb, Don Gibson, Kitty Wells, and Hank Williams Sr., who had a deep affection for Arthur Q. Smith.

After Smith had a stint at Acuff-Rose Music as a songwriter and at WMPS radio in Memphis, Hank Williams hired him as his business manager. His love of drinking and Hank's similar affection was a combination Audrey Williams would not tolerate, and she fired Arthur Q. Smith. It was his last real job. He returned to Knoxville, where he continued to write and sell songs to anyone who would buy them for drinking money.

In 1962, he was diagnosed with cancer and on March 22, 1963, was found dead in a flophouse at the corner of Vine Street and Summit Hill Avenue. He left behind a wife and six children and, unknown until then, countless receipts for songs he had sold. Songwriter Hall of Fame inductee Harlan Howard wrote the song "Be Careful Who You Love" as a tribute to Arthur Q. Smith. In 1990, none other than Hank Williams Jr., who had been tended as an infant by Arthur Q. Smith's wife, Lillian, released the song on his album *Pure Hank*.

Smith's face adorns a mural in Knoxville's Old City. He has been written about around the world and nominated for the Songwriter's Hall of Fame. The Pritchett children still reside in Knoxville and hope to preserve his legacy as a songwriter. His value to WNOX and to the field of country music, which first cast its net across Knoxville's airwaves, was representative of the incredible talent that stood behind its microphones and on its stages—making it among the most legendary radio stations in the United States.

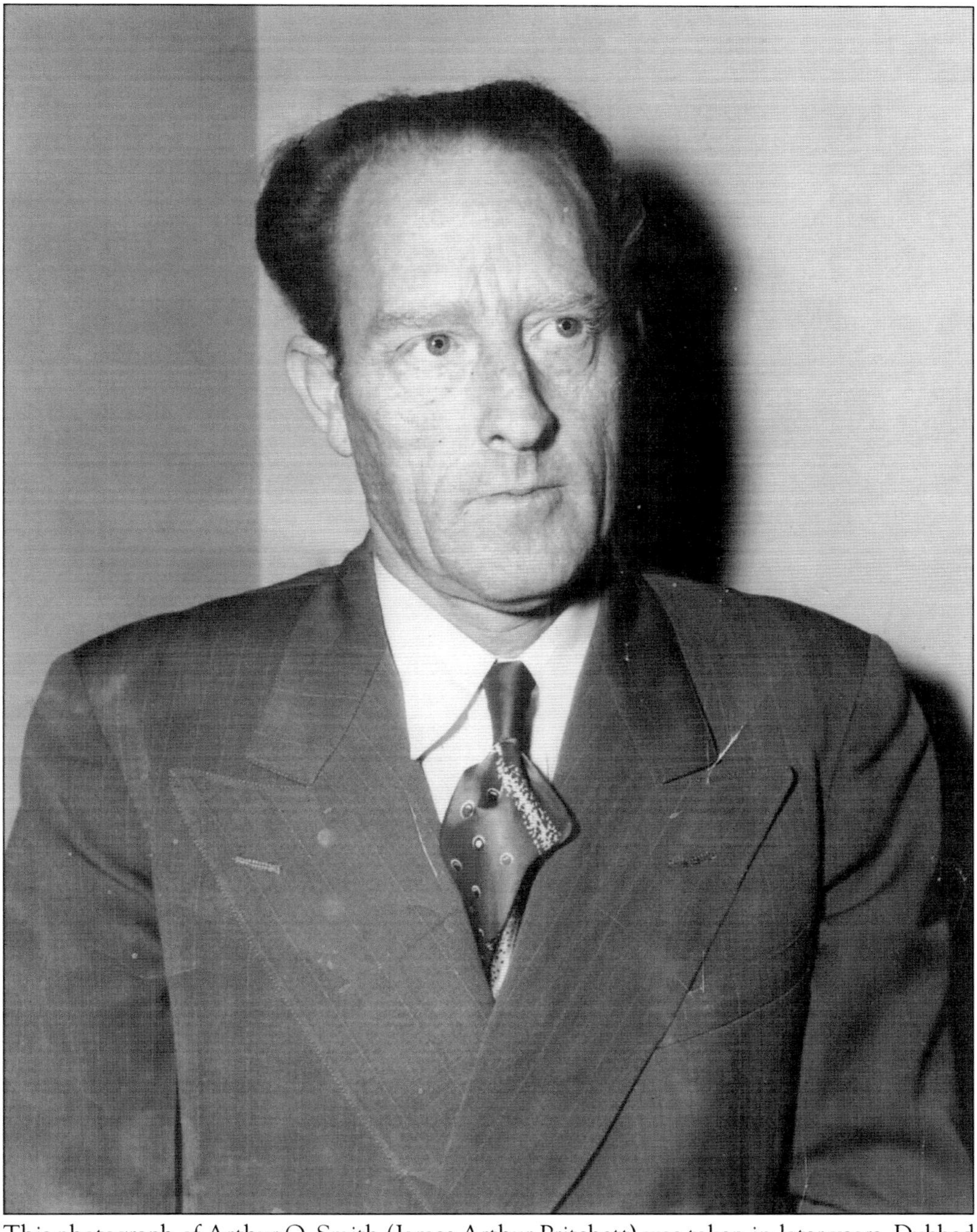

This photograph of Arthur Q. Smith (James Arthur Pritchett) was taken in later years. Dubbed the "Mystery Man of Country Music" by a London magazine in 2001, he had the ability to turn a lyrical phrase that was legendary among country music entertainers, but few people outside of the business knew of his remarkable talent. (Courtesy of Pritchett family.)

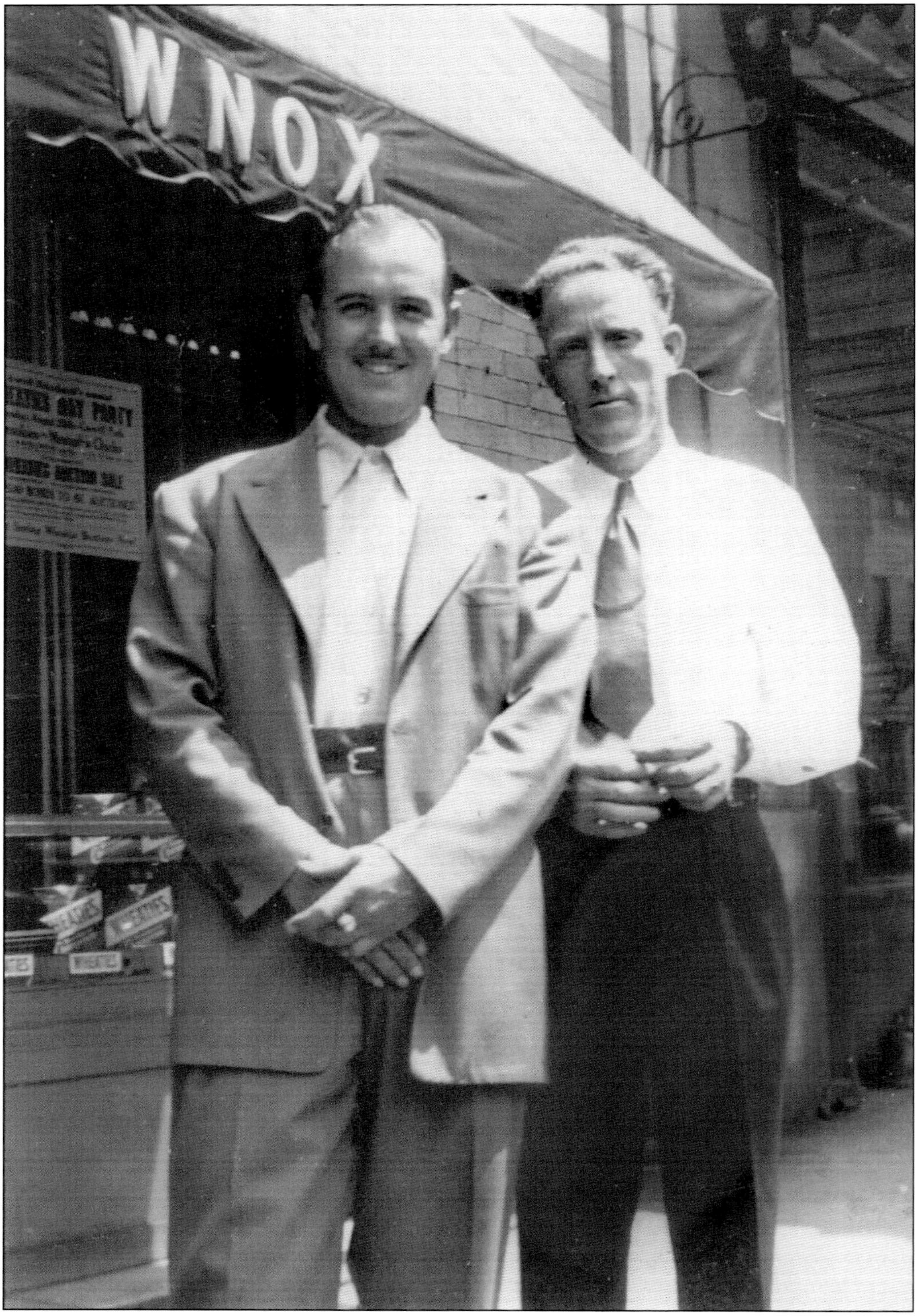

Arthur Q. Smith is pictured outside the WNOX studios on Gay Street with up-and-coming musician and comedian Archie Campbell (left). Campbell was a fixture of the Knoxville radio scene for many years in the 1950s, doing live comedy and singing. (Courtesy of Pritchett family.)

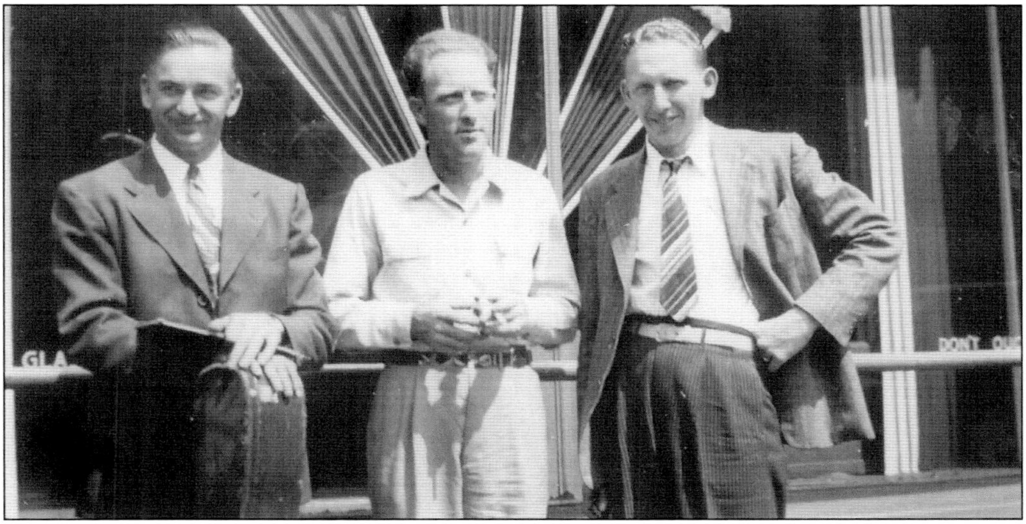

From left to right are musicians Curly Seckler, Arthur Q. Smith, and Leonard Stokes on Gay Street outside the WNOX studios. Seckler was an original "Foggy Mountain Boy" who played mandolin with Lester Flatt and Earl Scruggs on some of their greatest recordings. Previously, Scruggs lived in Knoxville while playing on WNOX's *Mid-Day Merry-Go-Round*. While known as a songwriter, Smith started out as a member of the WNOX house band and was always remembered as a much-loved and respected member of the band. (Courtesy of Pritchett family.)

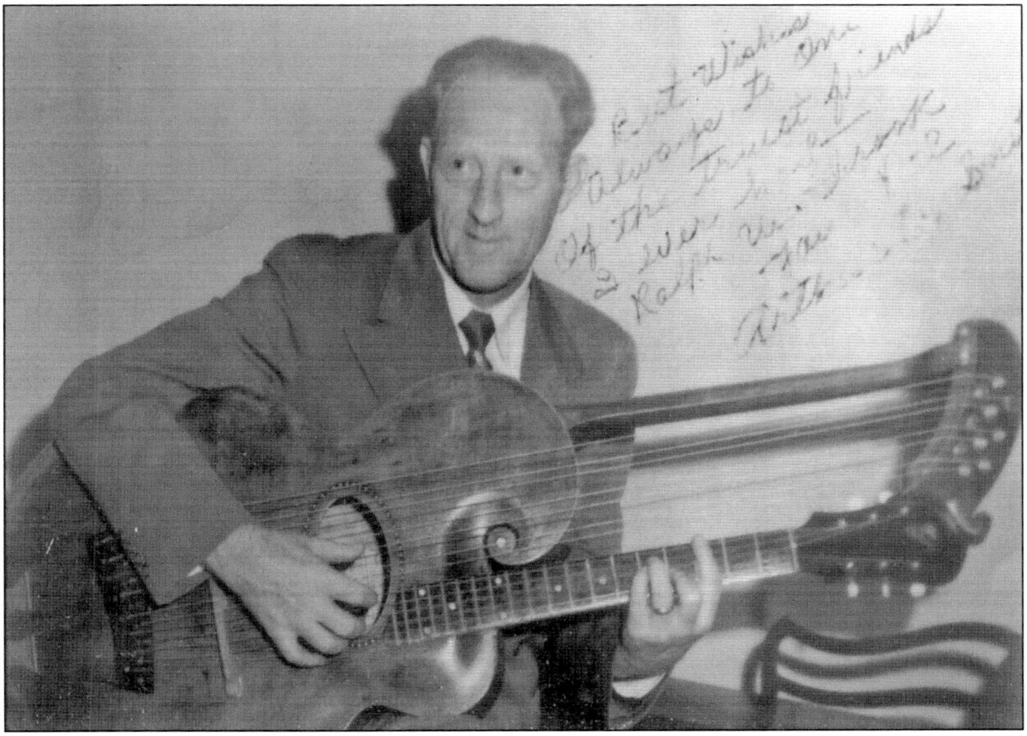

Smith is posed here with a harp guitar. This is the autographed photograph that seems to most often appear of Smith. The instrument extended the musical range of a guitar, and the unfretted strings were plucked like harp. For a songwriter, the added range gave a creative dimension not found on the normal six-stringed instrument. (Courtesy of Pritchett family.)

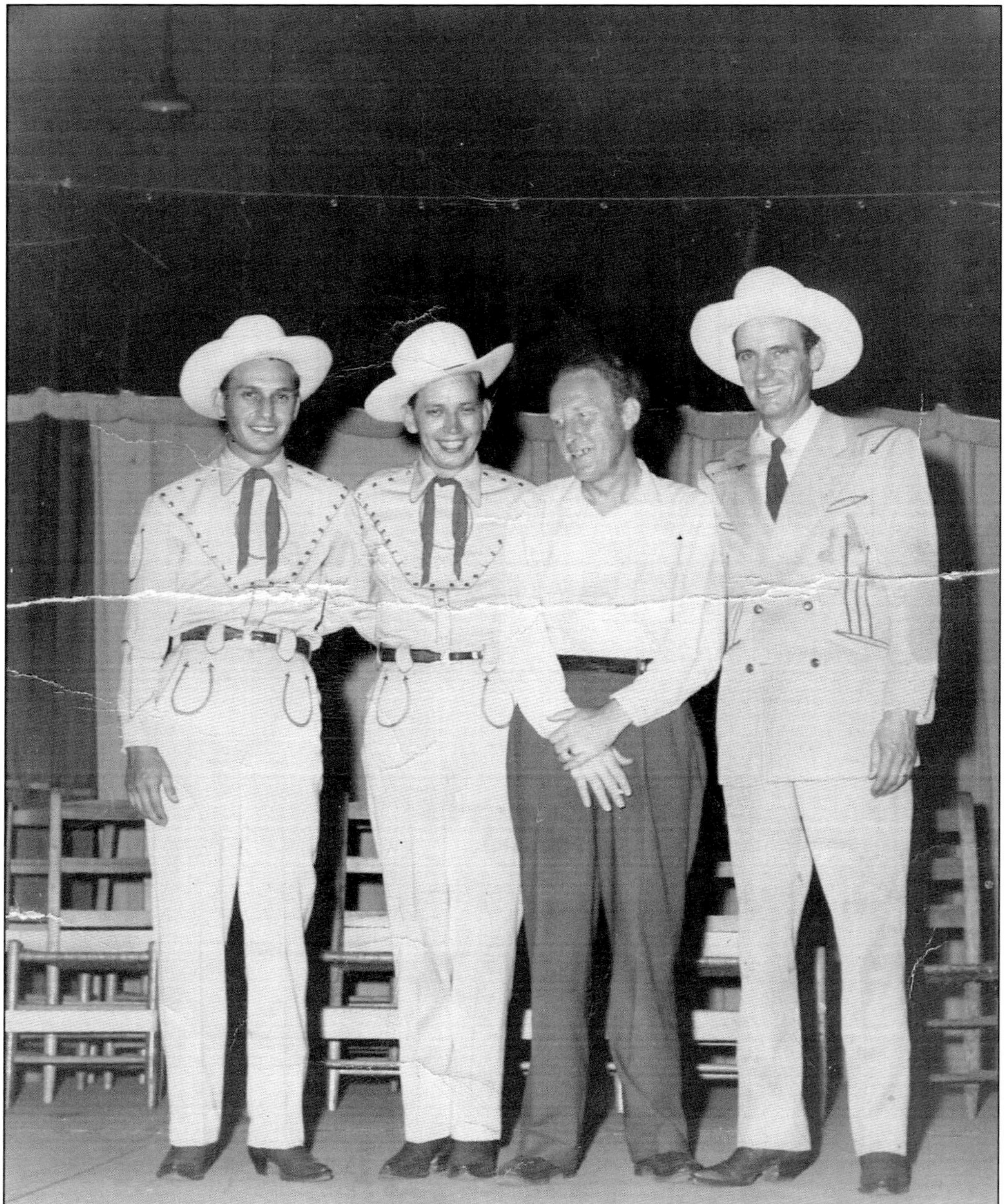

Arthur Q. Smith (third from left) and country music entertainer Ernest Tubb (far right) are pictured in a backstage photograph with the Texas Troubadours. Tubb appeared on WNOX in his early career and is known for being the first country music act to play Carnegie Hall. Arthur Q. Smith wrote the song "Rainbow At Midnight," which was a number one hit on Billboard Charts for Tubb in 1949, and "Missing In Action," which reached number three on the charts in 1952. His descendants still hold an album of unpublished songs, including the last song sold to Hank Williams. (Courtesy of Pritchett family.)

Two

Lowell Blanchard

Lowell Blanchard is a two-word name in country music broadcasting that is world renowned. He was born November 5, 1910, in Palmer, Illinois, the son of a grocer and the mayor of the city. He spent his summers working on the family farm and doing odd jobs. He got into radio while attending the University of Illinois and made a name for himself at the 1933 Chicago World's fair as a master of ceremonies. Blanchard wanted to be an actor but fell in love with broadcasting and emceeing. After Blanchard worked at radio stations in Indiana, Iowa, and Michigan, Dick Westergaard hired him into WNOX as an announcer and placed him in charge of local programming. WNOX was the only big game in town for Knoxville, but Scripps Howard marketing wanted to use the 10,000-watt broadcast signal to attract those outside the city in neighboring counties, which meant they had to find a way to appeal to their audience.

According to a Scripps Howard employee newsletter, Westergaard told Lowell Blanchard when he arrived, "You are to become a hillbilly. Bring us hillbilly performers. Entertain with them as one of them." Blanchard did just that. He drove into the hills and hollows of East Tennessee on rumors of a band or singer here and there. He developed an ear for the music and talked the shy, backwoods performers into coming on the radio and playing before an audience.

He created the legendary *Mid-Day Merry-Go-Round* and its Saturday night equivalent, the *Tennessee Barn Dance*. The entertainers started as a trickle and then became a torrent, pouring into the quiet little radio station at the top of the Andrew Johnson Hotel. There were never enough seats for everyone. People packed the hotel to see the shows, and management had to ask the station to leave. Blanchard was committed to the mission and overlooked nothing. He hired a guitar teacher, Chet Atkins, to play background on commercials and worked his unique sound onto the stage at WNOX. He is credited with bringing acts out of the hills and launching the careers of Roy Acuff, Archie Campbell, Kitty Wells, Chet Atkins, Mother Maybelle and the Carter Sisters, Pee Wee King, Martha Carson, Don Gibson, the Louvin Brothers, and many others who became giants in the country music industry.

Blanchard was not only selling WNOX, he was selling radio to people who were just discovering it. He invited them down to see the shows, and they came in large and unmanageable numbers. Blanchard—to Westergaard's dismay—was undeterred and bounced his programs around the city until the station settled in its longtime location on Gay Street.

Blanchard was a phenomenal voice on radio and an astounding master of ceremonies. He wrote his own scripts, did comedy, and mastered stagecraft. His good-natured approach and on-stage sales ability allowed him to put acts on the stage that could not have made it anywhere else, and it paid off big. He became the leading promotional voice of country music, and his fans numbered more than most bands who walked on WNOX's stages.

When the station changed its format to Top 40 in 1962, Blanchard remained with WNOX and became an emcee; it made him a great announcer for local sports, and he called the Knoxville Smokies minor league baseball games—making famous the on-air "Willhe" pitch on a full count: "Will he walk him, will he strike him out." His sports fans were as numerous as his program fans in the beginning days of country music. Blanchard also did on-air color for the University of Tennessee basketball team with legendary sports announcer John Ward. It was immediately following a basketball game on February 18, 1968, that WNOX's famous front man had a heart attack and died.

In April of that year, a who's who of country music entertainers and fans from across the nation descended on Knoxville's Civic Coliseum for a final *Mid-Day Merry-Go-Round* farewell honoring Lowell Blanchard's contributions to country music entertainment. Blanchard's talent went far beyond the radio station. He served two terms as a city councilman, unsuccessfully ran for mayor and state legislature, and donated his time to numerous community charities and regional events, such as the Cosby Ramp Festival, the Easter Seal Society, the March of Dimes, local horse shows, and the annual Tennessee Valley Agricultural and Industrial Fair.

He is remembered to this day as one of the founding fathers and promotional voices of country music. In 1977—nine years after his death—WNOX's Lowell Blanchard was inducted into the Country Radio Broadcasters Hall of Fame as one of the great announcers of all time, where he was a role model for three generations of broadcasters and regarded as the first genuine on-air-voice of East Tennessee.

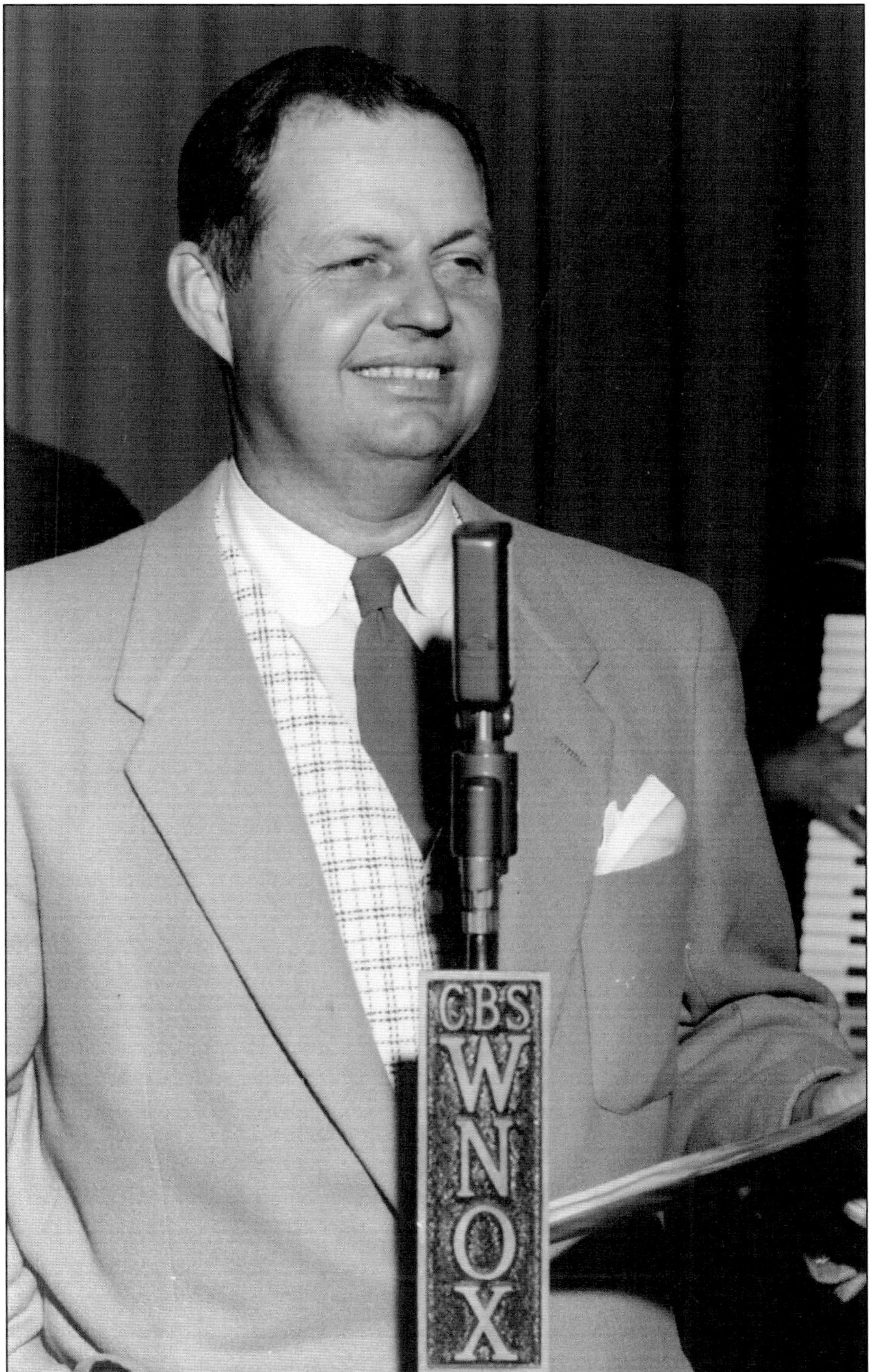

This is the staff photograph of Lowell Blanchard that was used when he was nominated into the Country Radio Broadcasters Hall of Fame. His work discovering and promoting talent from the hills and valleys of the East Tennessee region has never been equaled.

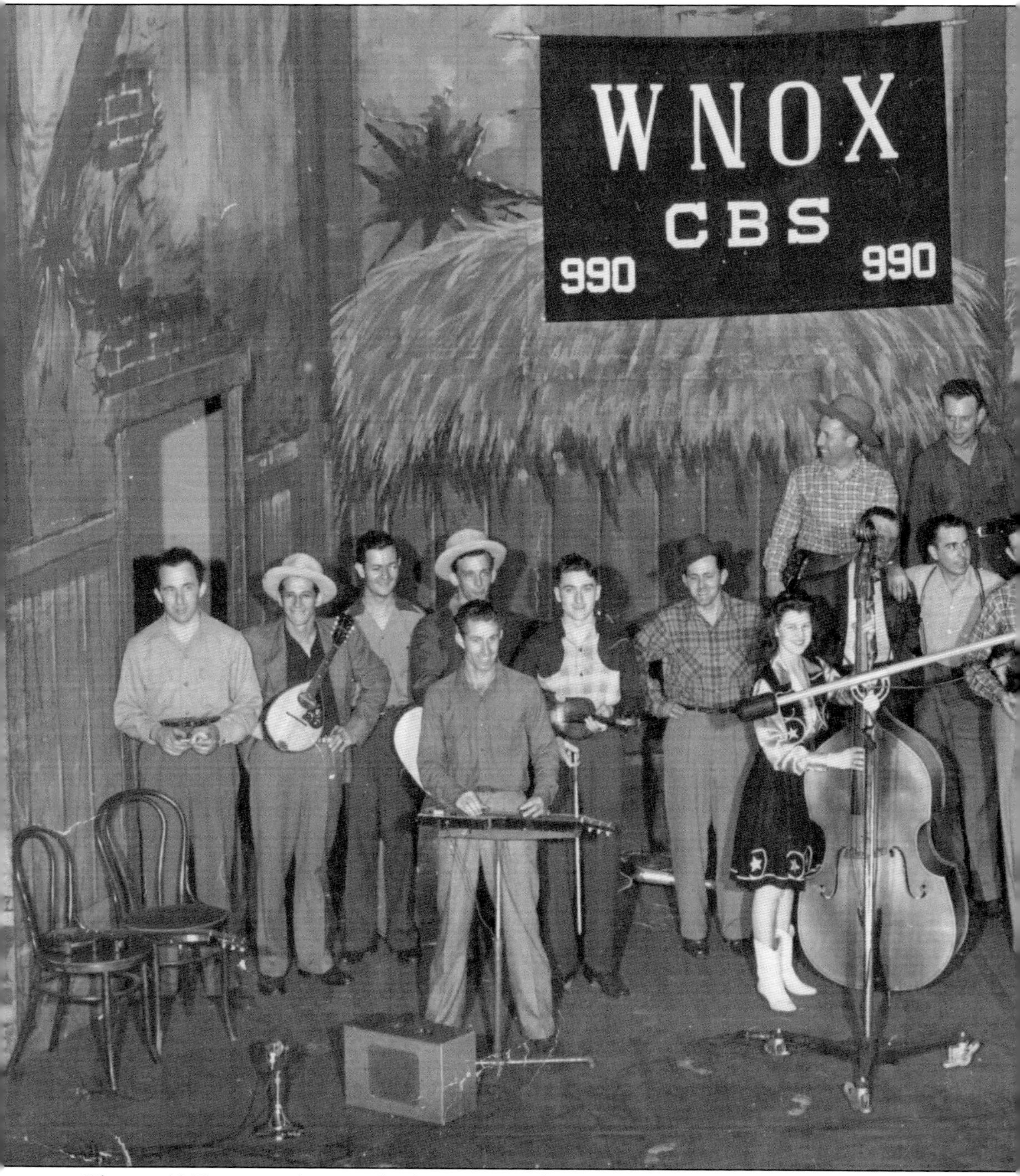

This is a 1947 vintage photograph of WNOX's *Mid-Day Merry-Go-Round* at one of its high points with performers whom Lowell Blanchard was featuring on the daily radio program. While many of them on stage are unidentified, those who are include John Gallaher (far left), Joe Burchfield (second from left), Bill Burchfield (fourth from left), Wilbur Manning (fifth from left), Lester Flatt (seventh from left), Archie Campbell (eight from left in back), and Sonny James Loden (12th from left in front), who dropped his last name and recorded as Sonny James in Nashville. Loden would go on to be one of country music's greatest stars and the first country music entertainer

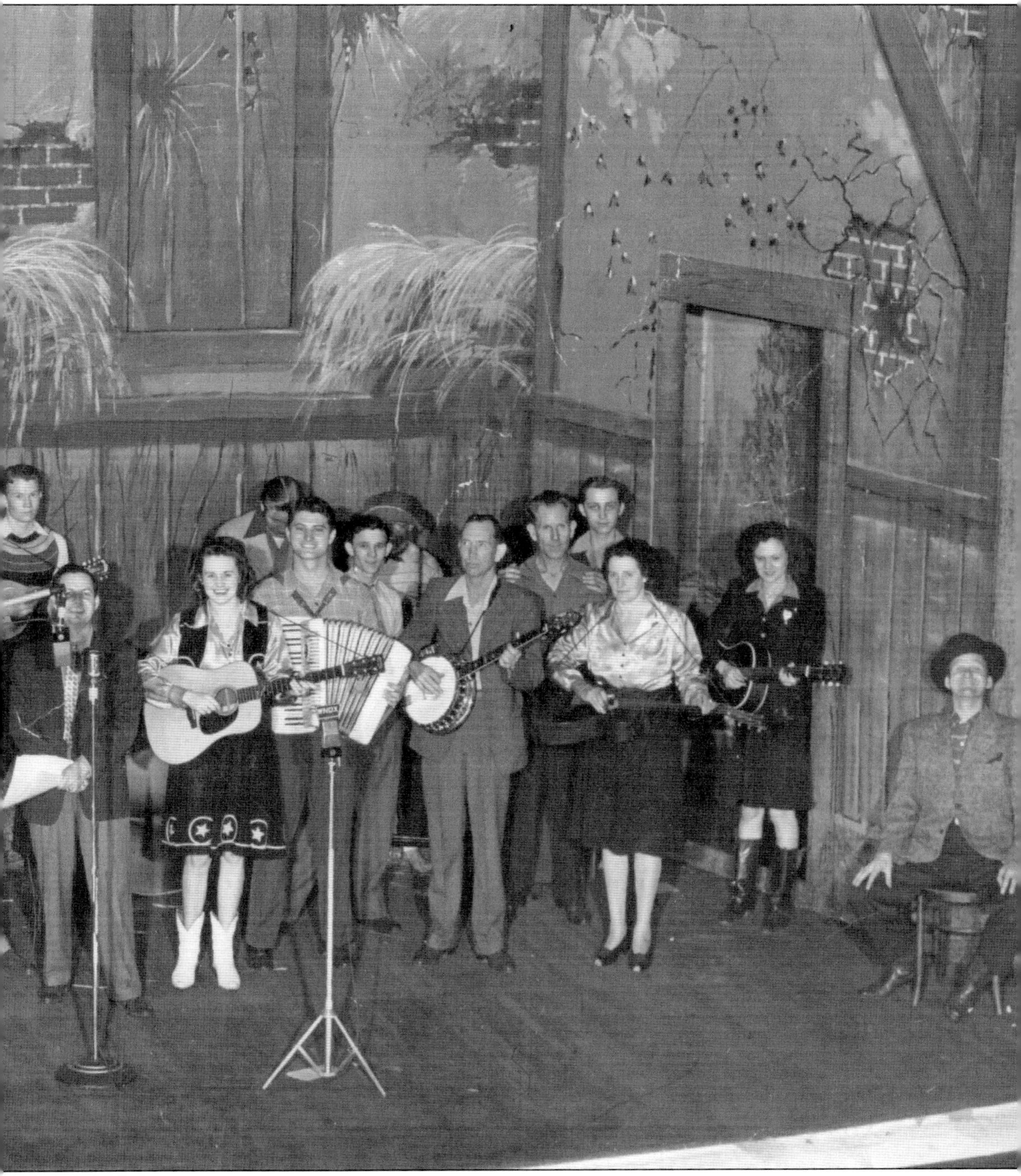

to be honored with a star on the Hollywood Walk of Fame. Lowell Blanchard is center stage behind the microphone. To his right is the rest of the Loden family, who performed as regulars on the program. Buddy Banes is on the accordion, and the woman at far right with the guitar is identified as Kitty Wells, who was also a regular on the *Merry-Go-Round* before launching into a stellar country music career in Nashville. Charles Elza ("Kentucky Slim") is sitting on the stool. (Courtesy of Patsy Robertson.)

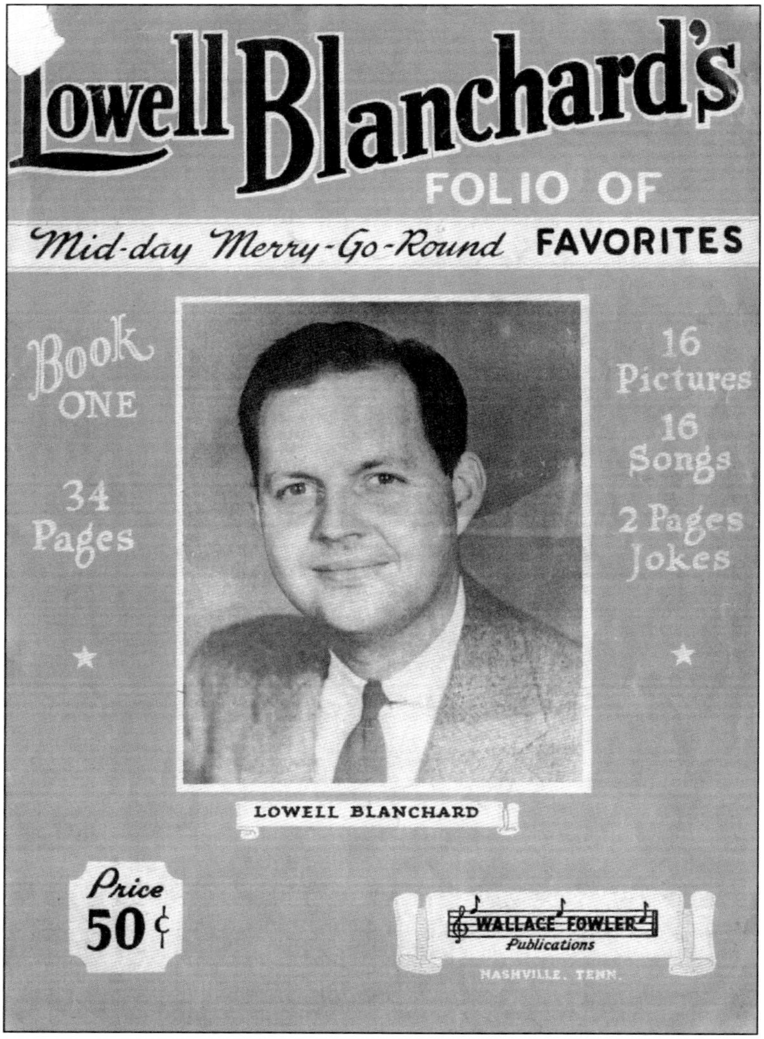

Blanchard earned his own following as an onstage personality as well as a broadcaster. In 1945, he released a collection of songs and jokes he used on stage to entertain the crowds between sets. The songbook is a quasi-modern collection of folk tunes written in post–World War II America, which included such songs as "When Atoms Fell on Hiroshimo" [*sic*], and on-stage bits used to get a laugh.

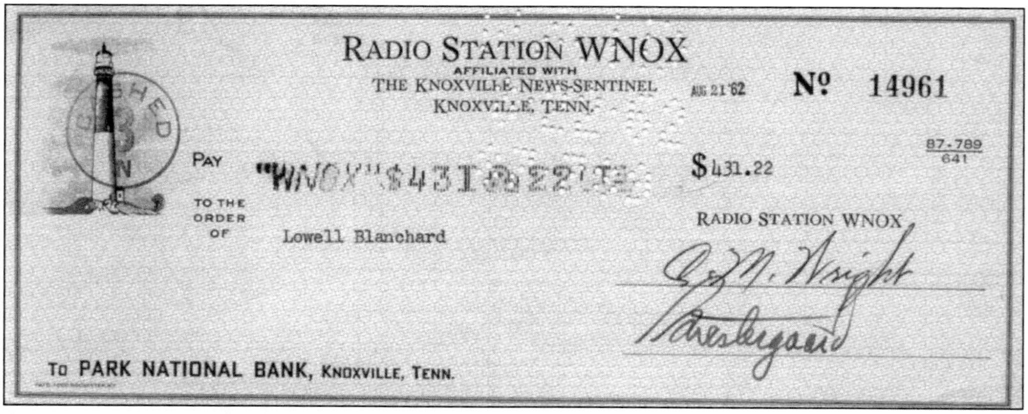

A 1962 WNOX paycheck to Lowell Blanchard showing the familiar Scripps Howard logo reveals a more modest sum than most would expect for such a legend. It was a seminal year for WNOX as the station moved from its country music format to Top 40.

Lowell Blanchard's professional career continued to evolve as radio took root and moved forward. He still served as master of ceremonies of the *Tennessee Barn Dance* and other radio-sponsored programs, but he was also a well-known local sports broadcaster who provided color. He is pictured here (standing) with legendary "Voice of the Volunteers" announcer John Ward. (Courtesy of University of Tennessee Athletic Department.)

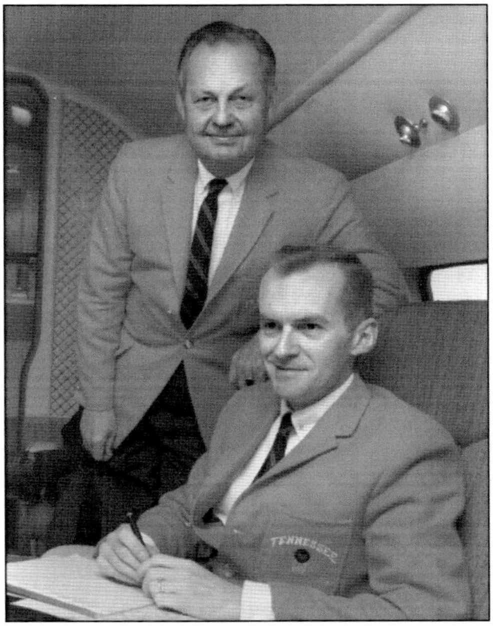

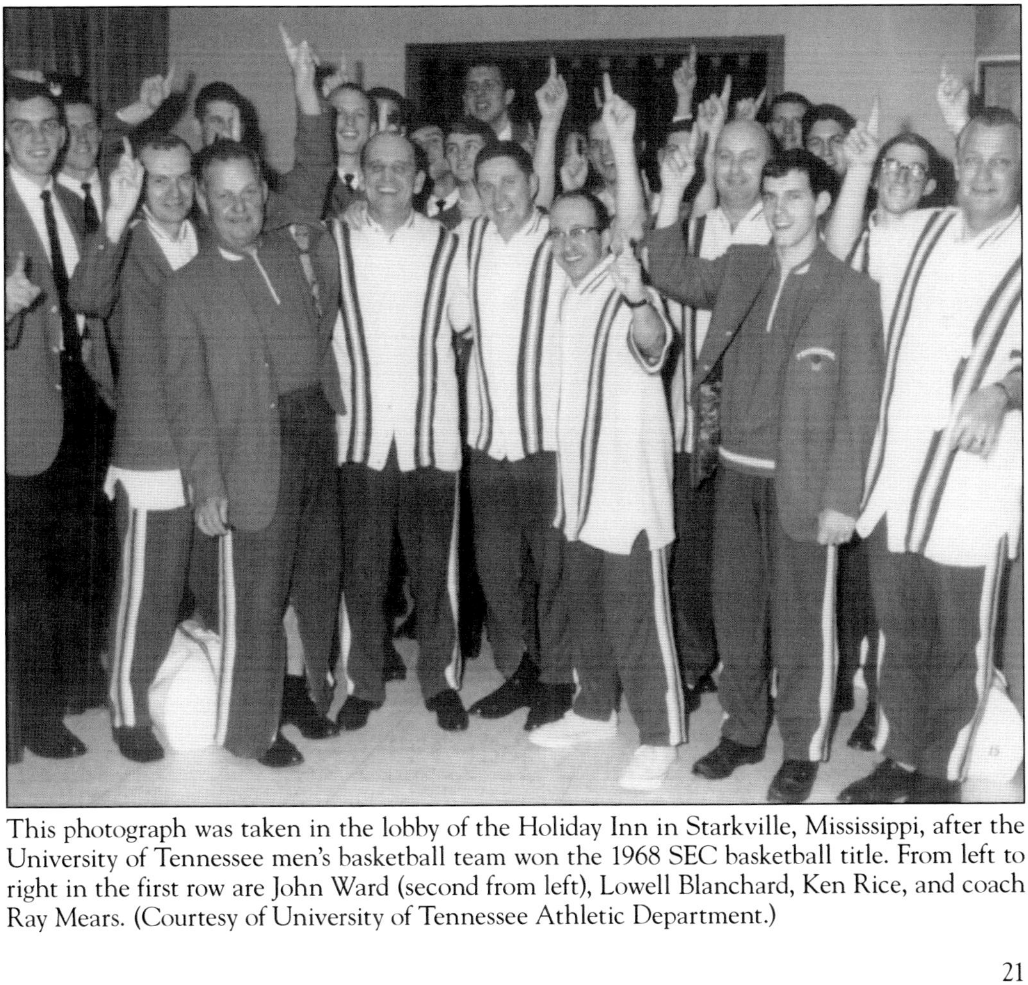

This photograph was taken in the lobby of the Holiday Inn in Starkville, Mississippi, after the University of Tennessee men's basketball team won the 1968 SEC basketball title. From left to right in the first row are John Ward (second from left), Lowell Blanchard, Ken Rice, and coach Ray Mears. (Courtesy of University of Tennessee Athletic Department.)

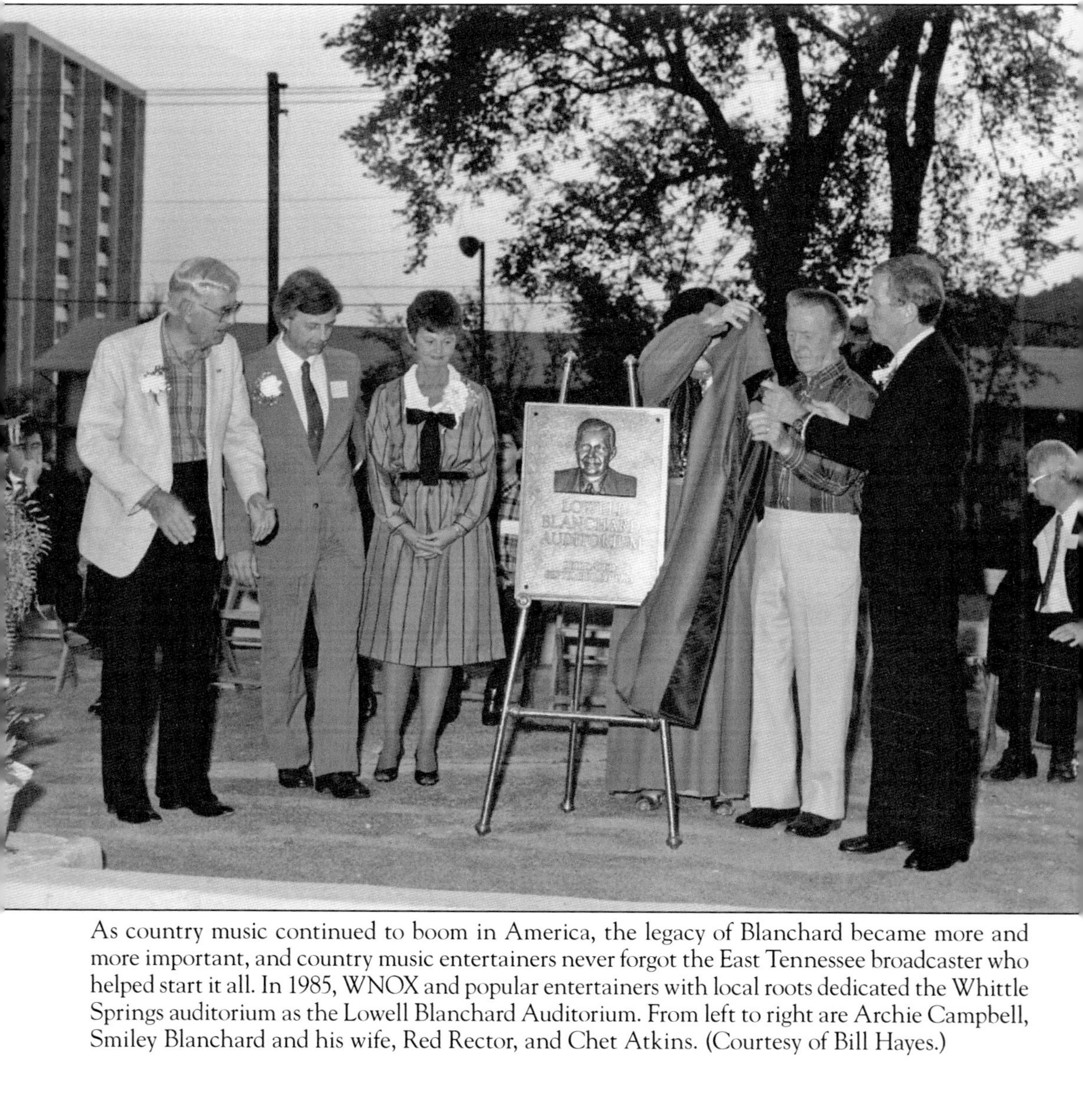

As country music continued to boom in America, the legacy of Blanchard became more and more important, and country music entertainers never forgot the East Tennessee broadcaster who helped start it all. In 1985, WNOX and popular entertainers with local roots dedicated the Whittle Springs auditorium as the Lowell Blanchard Auditorium. From left to right are Archie Campbell, Smiley Blanchard and his wife, Red Rector, and Chet Atkins. (Courtesy of Bill Hayes.)

Three

The Scripps Howard Era

WNOX AM was purchased by Scripps Howard on December 17, 1935. The newspaper and magazine company had bought WCPO in Cincinnati earlier that September and was branching into a new form of mass media cautiously. The idea of transferring a full-service newspaper mentality to broadcasting was an experimental concept they took seriously. They already owned the *Knoxville News Sentinel* and had an invaluable insight into the East Tennessee region few people outside the hills and valleys understood.

They were not as worried about necessarily capturing the ear of Knoxville as they were the people and families that visited the city once a week or so to do business, shop, and buy groceries. It represented a great untouched part of the population their newspaper writers had been covering for years. With a 10,000-watt signal booming from downtown and radios becoming ever more popular as electricity began powering up the region, it was a rare ground-floor opportunity to be the first to reach them, and WNOX's parent was not going to miss it.

Richard Westergaard was sent in as general manager to take over the station. He brought in Lowell Blanchard, hijacked O. L. Smith from the *News Sentinel*'s national advertising department, and assembled a team of people dedicated to reshaping the Knoxville radio station into an East Tennessee institution; for Scripps Howard, it was all about community. As with newspapers of the day, radio was expected to inform, educate, enlighten, and entertain its listeners. News, information, local programs, and national events suddenly beamed through the airwaves, and the sheer influential power of radio on all the masses—not just the literate—was immediately felt.

Lowell Blanchard was the prize for the station. It helped that both he and Westergaard were prolific promoters, but to sell radio to people, they invited them in to not only see the magic of it all but to be a part of it. Blanchard brought them in by the truckloads. The accommodating staff of the Andrew Johnson Hotel had grown really tired really quickly of the hundreds of people who came to see it, and Blanchard was not going to let up for a second. Westergaard became concerned. It was going to be difficult to sell WNOX as an asset to the city's business community if they were hurting the business of Knoxville's premier hotel.

Scripps Howard executives were simply astounded at the growing response and gave Westergaard the nod to move the station from the Andrew Johnson Hotel and lease the building at 110 South Gay Street. He grabbed buckets of red barn paint, and he along with anyone standing nearby, including Scripps Howard executives Jack Howard and Jim Hanrahan, started painting the walls of the new radio station while carpenters and electricians worked to get the new facility in order. Red has been a part of WNOX's logo ever since. Westergaard was quickly learning how East Tennesseans did things as well. He had ample funds available to put seats in the street-level auditorium, but the crowds were so massive, he placed pews instead so they could always "squeeze" one more into the row.

Scripps Howard and WNOX always stayed on the forefront and in step with their listeners on local news and events that were important to East Tennessee in particular. Agriculture was and remains the state's biggest industry. The dust bowls of the West in the 1930s illustrated how important it was to keep family farms up to date with new agricultural techniques or face crashes similar to ones on Wall Street that devastated far more than the economy. Farming brought in much needed capital to East Tennessee, but its impact was felt far from the fertile hills and valleys of the region. WNOX staffed a farm director, who broadcast in the early morning hours when workers were tending livestock or getting ready to start their day. Its influence on the region's airwaves cannot be overstated. It tied a family farmer industry together and informed one of what the other was doing to get better yields from crops or how to recycle land.

Scripps Howard and WNOX literally served the community and defined themselves by it. A city nestled in the hills once pleasantly starved for entertainment suddenly came to life with a vibrant musical community. Little did the townsfolk know the guitar players and hillbilly bands walking by them in the early days would go on to be some of the greatest country music entertainers in the industry, but it was still only one aspect of WNOX broadcasts that brought the world to East Tennessee. Radio was block programming then and a full-service enterprise. It was a mix of talk and musical variety. A listener tuning in would hear operas, symphonies, drama, world news, national news, live coverage of breaking events, and the programs and broadcasts that made it a principal information resource for East Tennessee. Being the first station to sign on the air in Tennessee did not make WNOX the legend it became; it was the people behind the microphone who did.

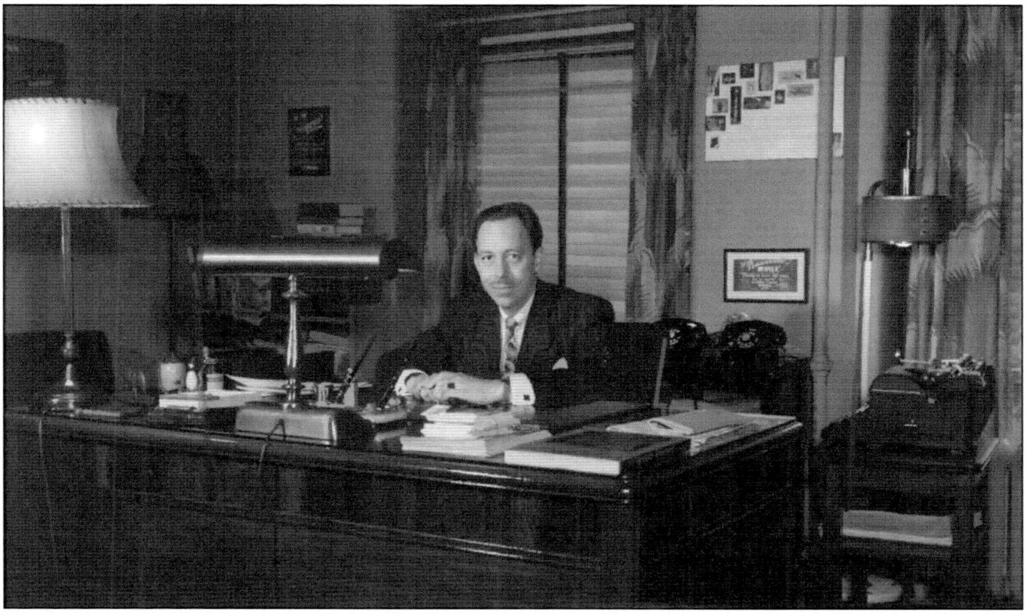

Richard "Dick" Westergaard (above) started work at a Des Moines, Iowa, newspaper and worked his way to columnist before starting work at a local radio station, helping promote the new medium. Jim Hanrahan, who had managed WNOX under Liberty Life in Knoxville before joining Scripps Howard, made the call to recruit him to the Knoxville radio station as general manager. (Courtesy of E. W. Scripps Company.)

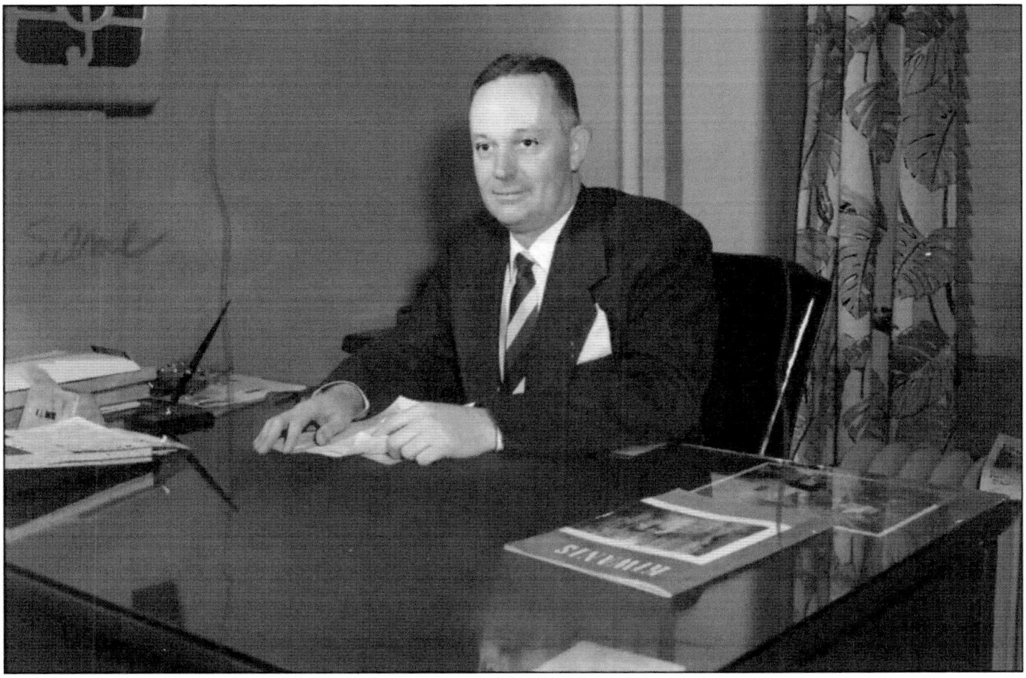

O. L. (Oliver Lee) Smith was working in the Scripps Howard–owned *Knoxville News Sentinel* as a sales executive in the national division when Westergaard hired him to WNOX. He later served as station director, ran the operation when Westergaard was serving in World War II, and eventually succeeded Westergaard as general manager. (Courtesy of E. W. Scripps Company.)

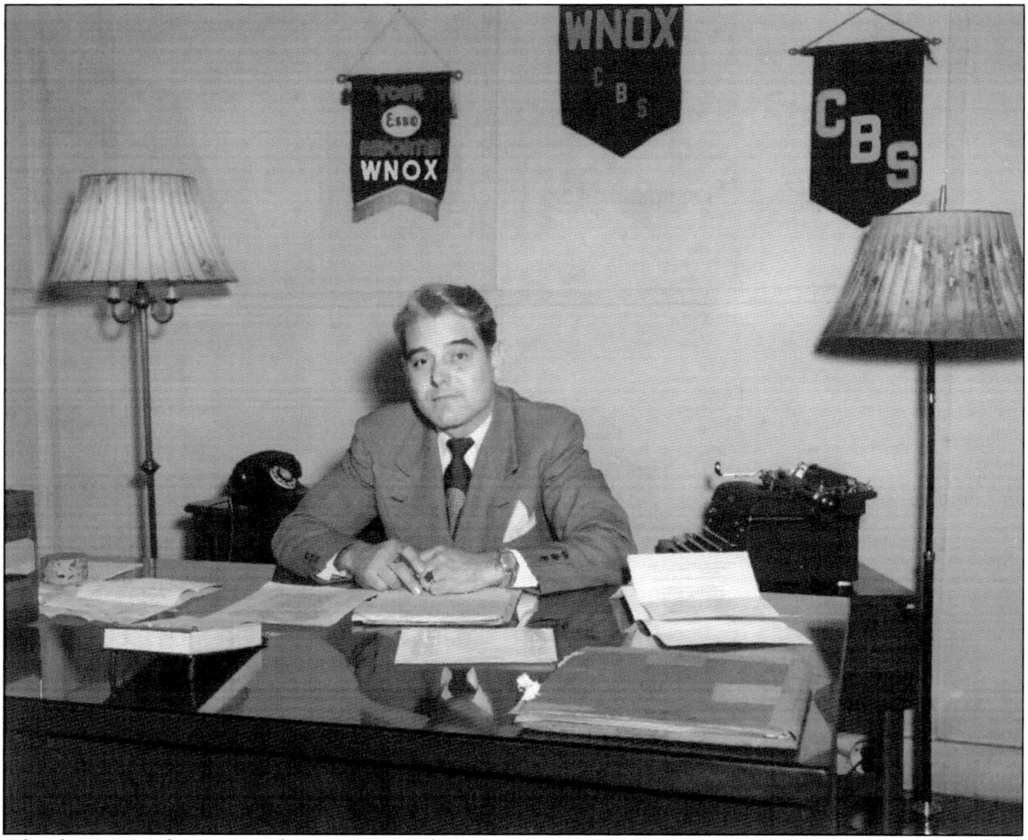

Charles McMahon served as program director at WNOX. While each WNOX employee had designated titles and duties to perform, each was also a part of the radio on-air staff and cross-trained to perform other jobs. (Courtesy of E. W. Scripps Company.)

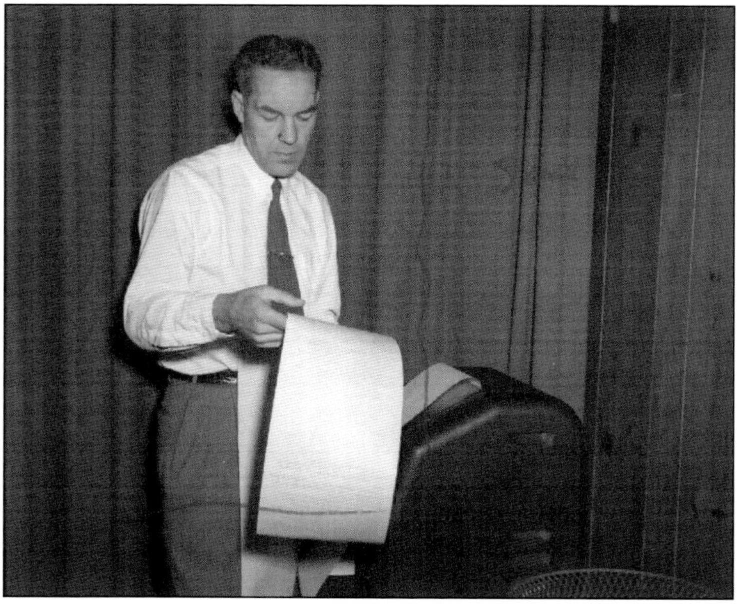

Farm director Cliff Adams stands at the United Press International (UPI) ticker pulling news feeds. He was responsible for researching and creating the farm programs and establishing its broadcast schedule for maximum effect. They were among the most-listened-to programs in the rural regions of East Tennessee. (Courtesy of E. W. Scripps Company.)

Robert Martin was a producer at WNOX and one of its announcers. He was known for recording tracks of himself singing different parts and playing it back as a duet—a trick learned as a Broadway actor. He was from New York City and produced *20 Questions* before making the move to Knoxville. (Courtesy of E. W. Scripps Company.)

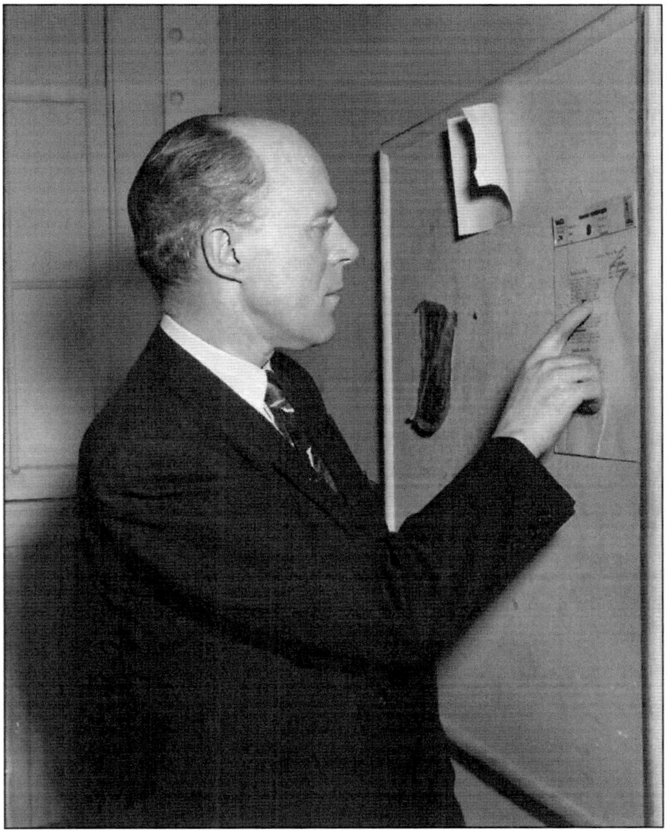

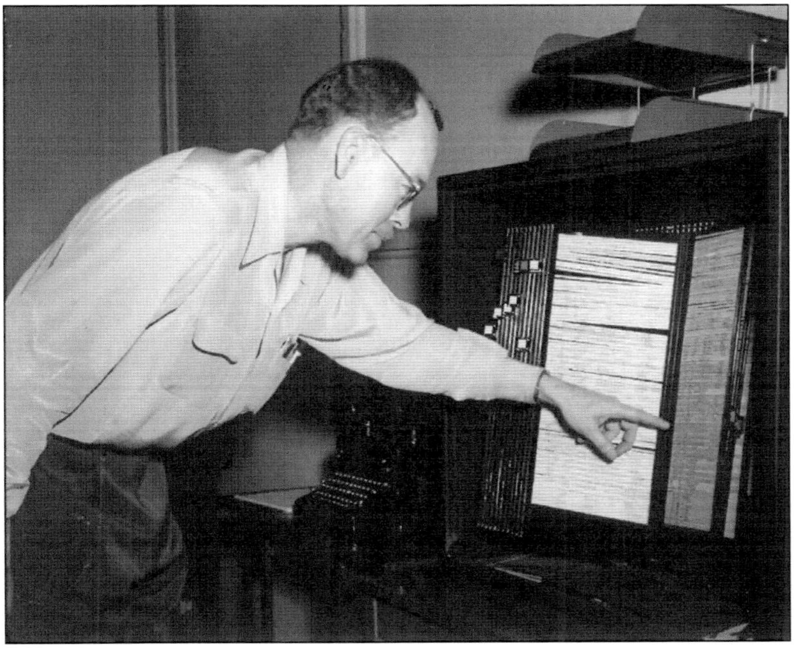

Hubert Carter was traffic director at WNOX but best known as a bass player for the *Mid-Day Merry-Go-Round* who served as the straight man for many of Lowell Blanchard's on-stage routines. (Courtesy of E. W. Scripps Company.)

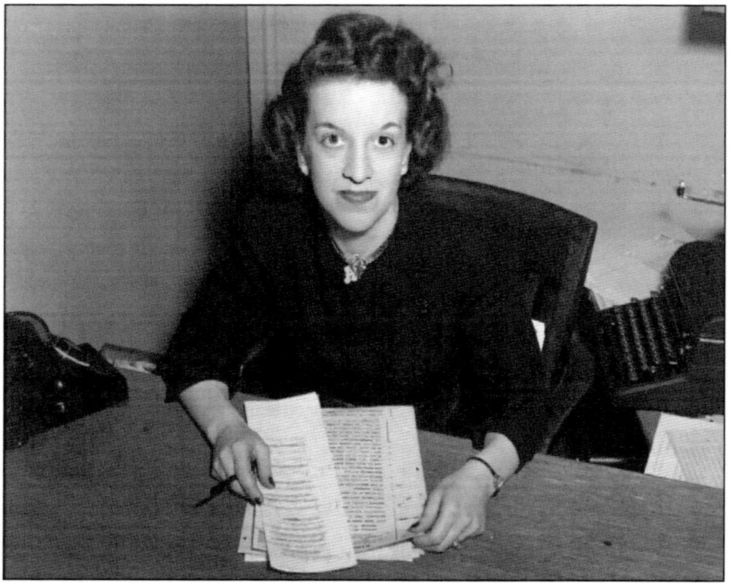

Caroline Seepe was a "continuity director" at WNOX. She would write commercials for the station and process them for recording by the station's various voices. (Courtesy of E. W. Scripps Company.)

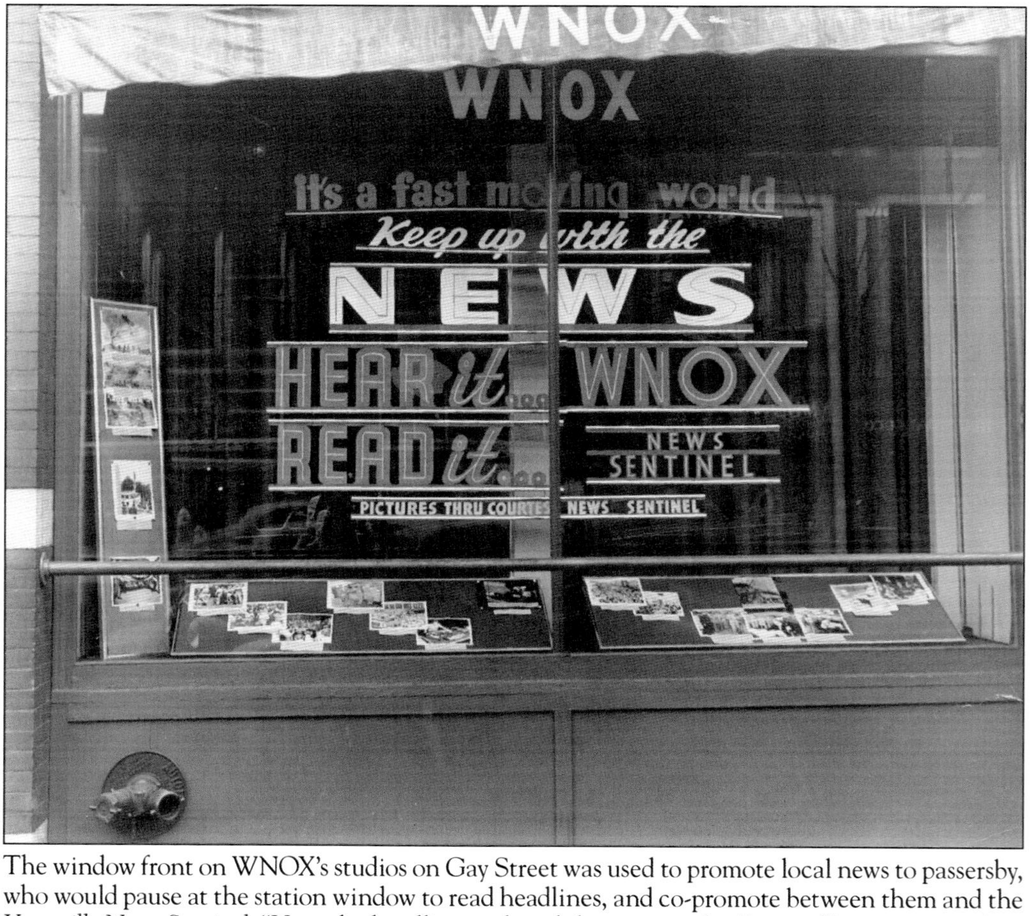

The window front on WNOX's studios on Gay Street was used to promote local news to passersby, who would pause at the station window to read headlines, and co-promote between them and the *Knoxville News Sentinel.* "Hear the headline and read the story in the Sentinel" was a resounding success for both organizations.

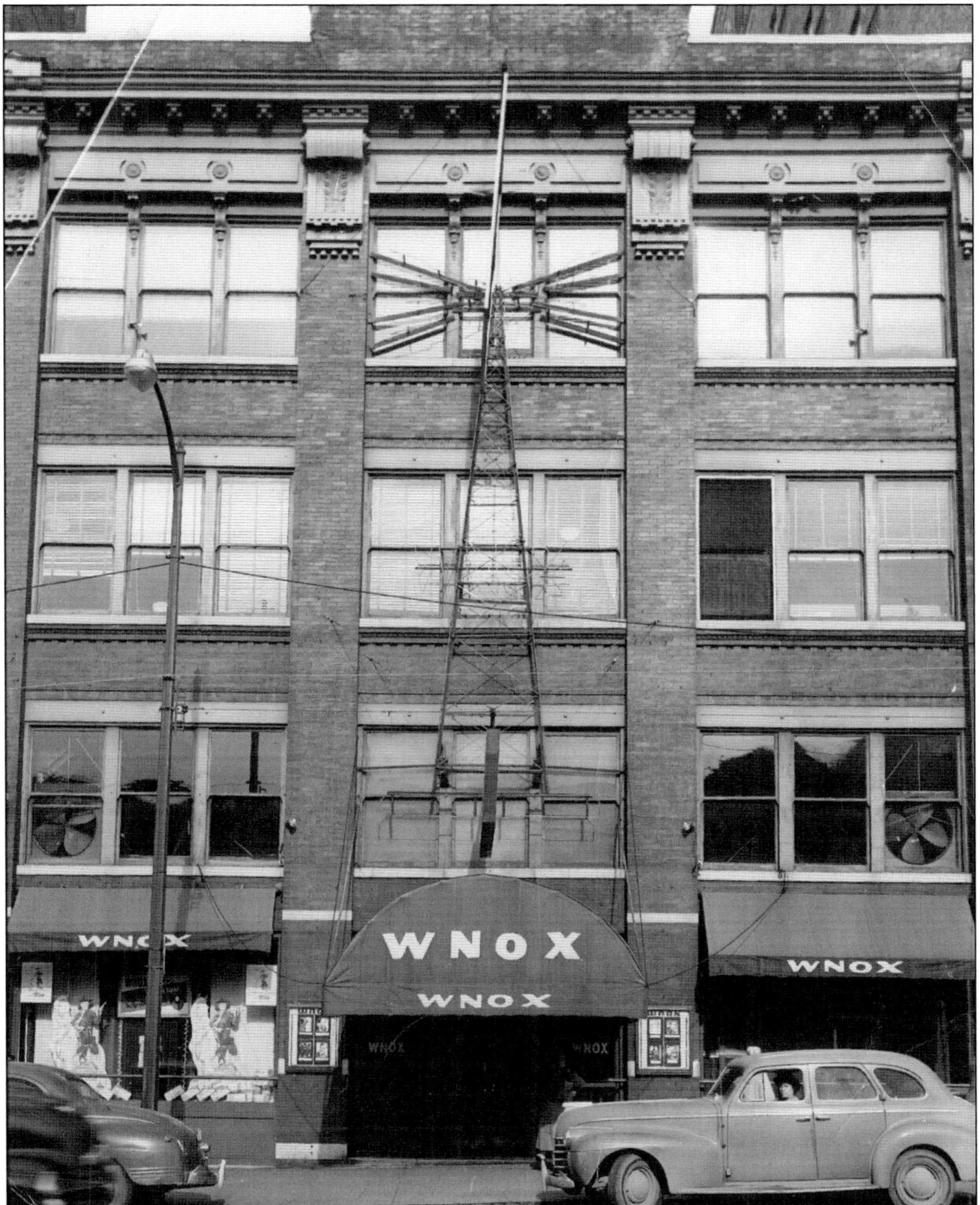

This is a photograph of WNOX's Gay Street office. The tower would light up at night to attract pedestrians. The station broadcast at 10,000 watts and pumped a signal across the Eastern Seaboard with the uncluttered airwaves and natural strength of the AM signal. Chief engineer Ray Evans is considered one of the unsung masterminds in the station's history, responsible for broadcasts from the studio and remote broadcasts from locations throughout Knoxville.

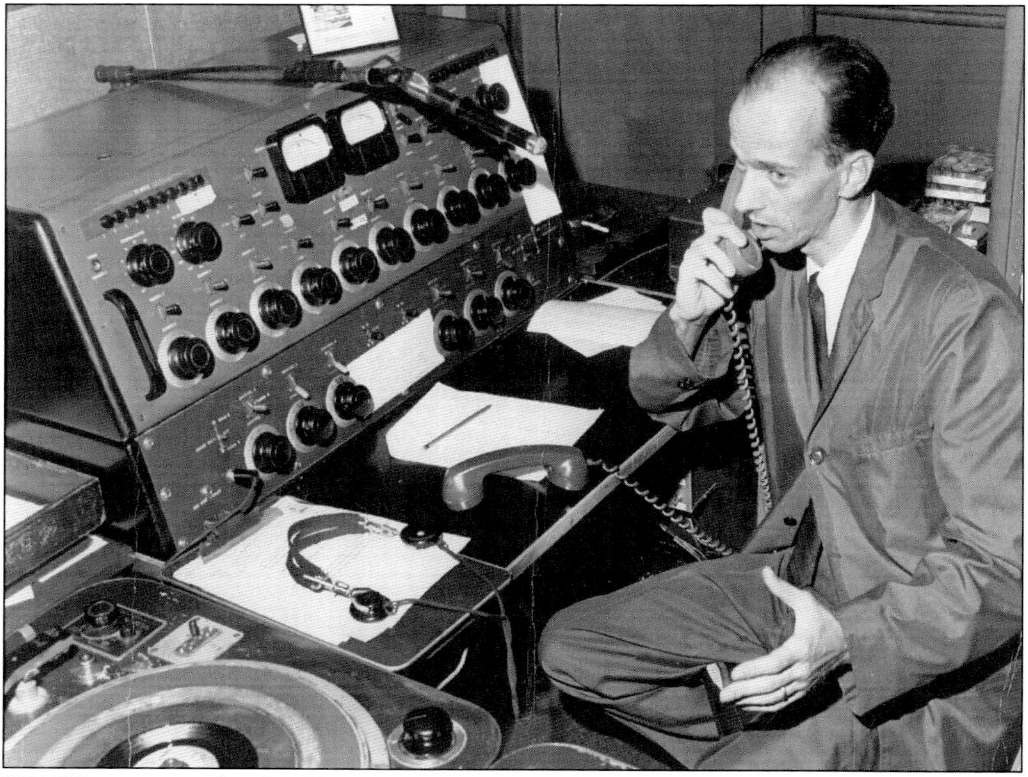

WNOX announcer Jim Small answers the hotline during a shift at the radio station. Announcers did a variety of tasks while on air and made sure programs and music maintained a seamless broadcast for listeners.

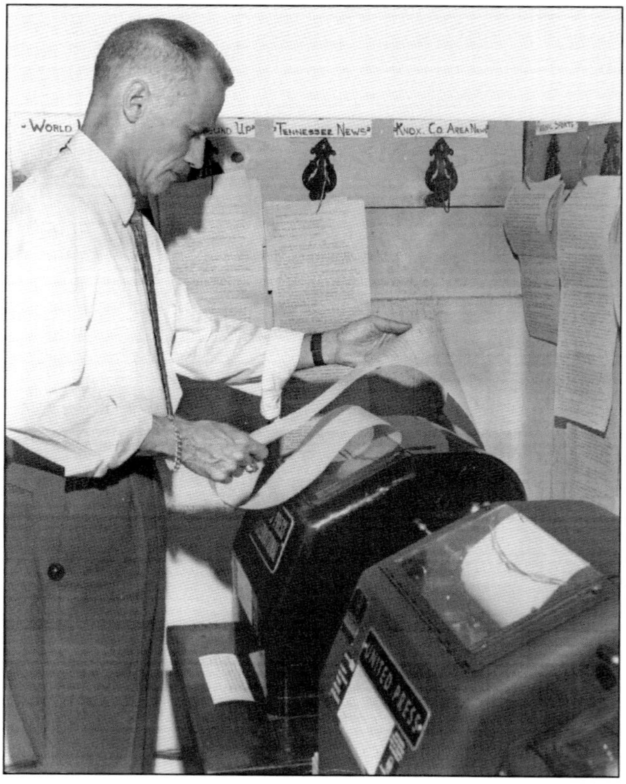

WNOX news director Ron Ashburn was a longtime employee of WNOX. Here he stands at the UPI wire tickers pulling stories for a news broadcast. Scripps Howard had objected to Associated Press's early business model, which selected one paper in a city to get its national feeds and stymied other newspaper startups. Scripps Howard created UPI as a news wire and made it available to everyone. WNOX was the first radio station to employ the service.

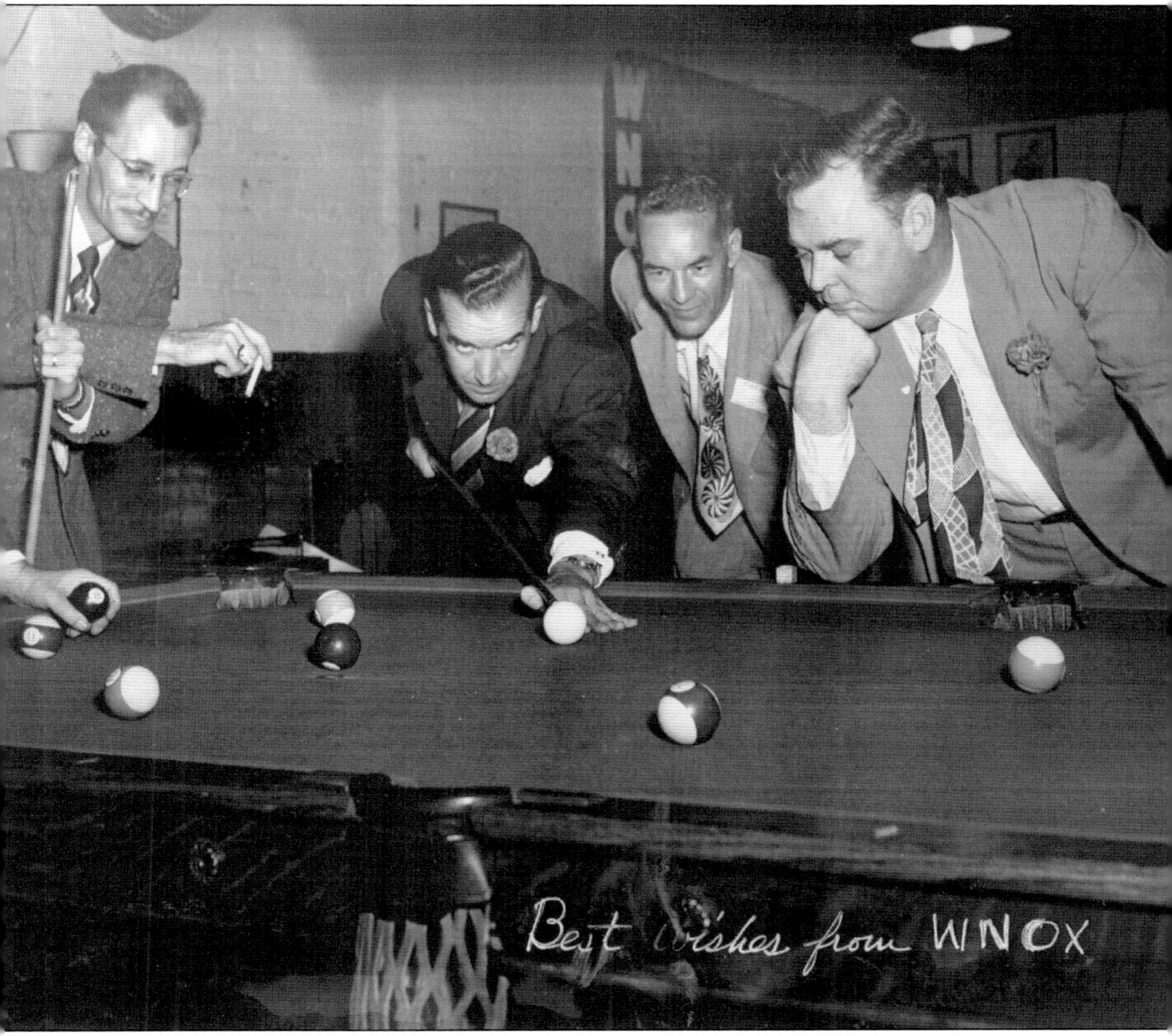

Here is a classic shot from the WNOX lounge upstairs at the Gay Street studio. The lounge was started with a $1,000 Dupont Award check won by Para Lee Bell for her broadcast reports on air pollution in Knoxville. From left to right are Fred Shepherd's hand holding the eight ball, Ron Ashburn, CBS National News anchor Lowell Thomas, advertising executive Jack Rowe, and station director O. L. Smith. Among one its most noted broadcast news credits was WNOX's first reports inside the then-secret city of Oak Ridge.

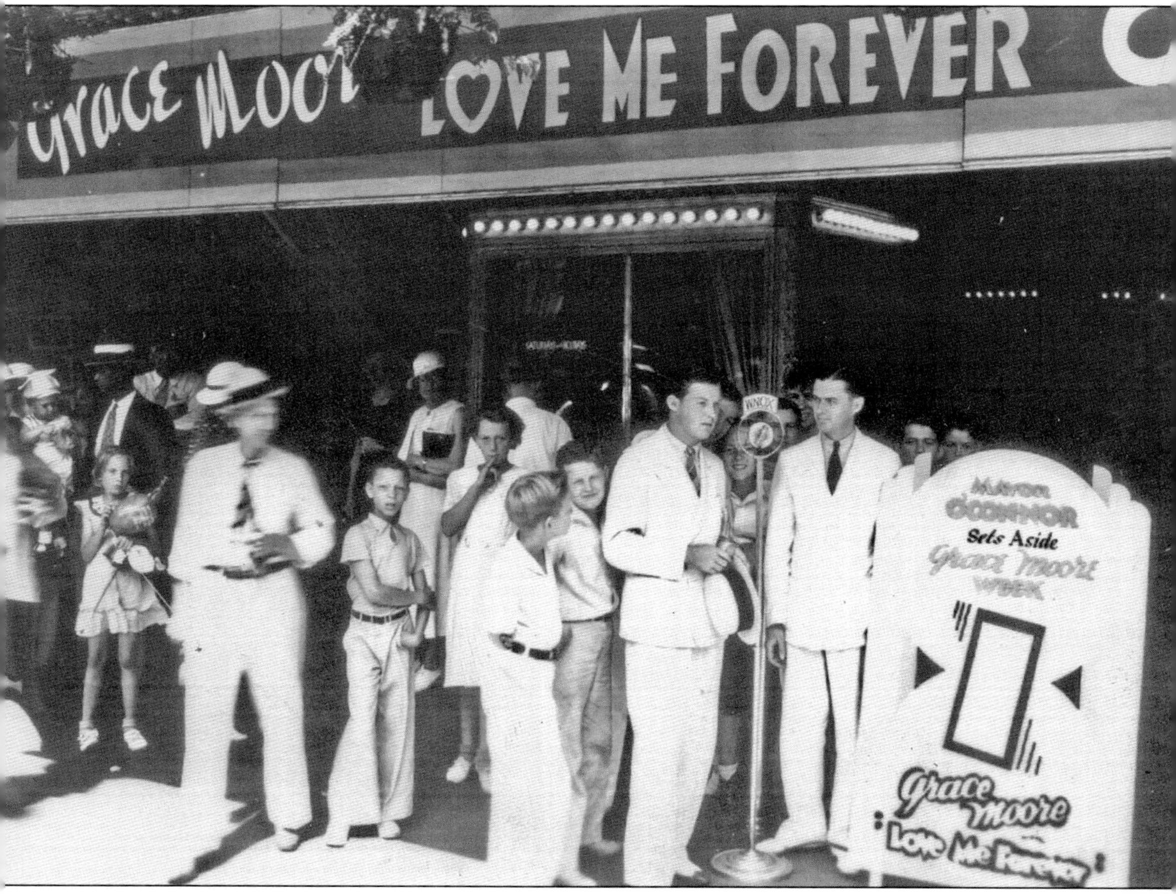

Advertising executive Fred Shepherd was one of the longest-serving employees of WNOX. He started as a board operator in 1931 while the station was still owned by Liberty Life, transitioned in the same job when Scripps Howard bought the station, and began his rise through the ranks. Here he is seen on the left of the WNOX microphone outside a theater. Shepherd started the station's "man on the street" interview that went on to become one of Lowell Blanchard's trademarks for the station.

World War II interrupted the careers of many talented employees at WNOX. Shepherd, Westergaard, and others were drafted into the war effort. This photograph was taken of Fred Shepherd (second from left) wearing the uniform in Miami Beach. Here he sings with two unidentified musicians for WKAT radio.

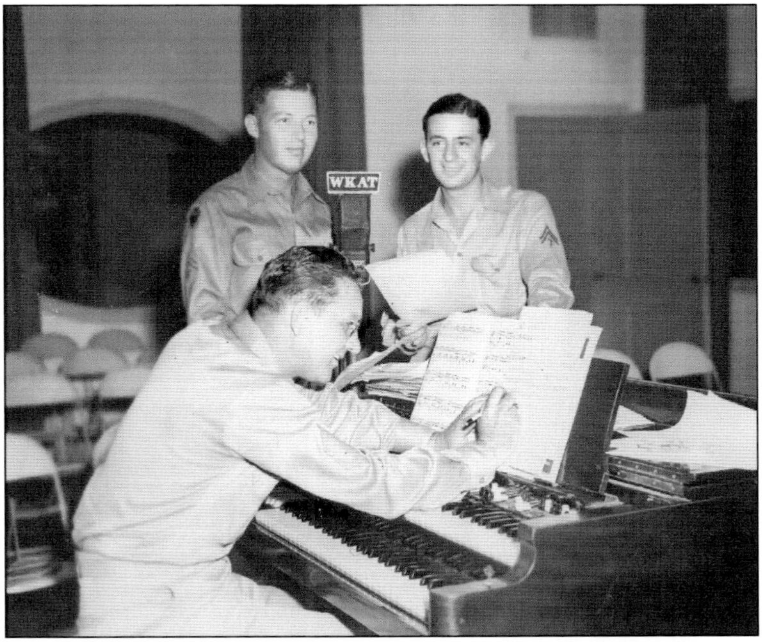

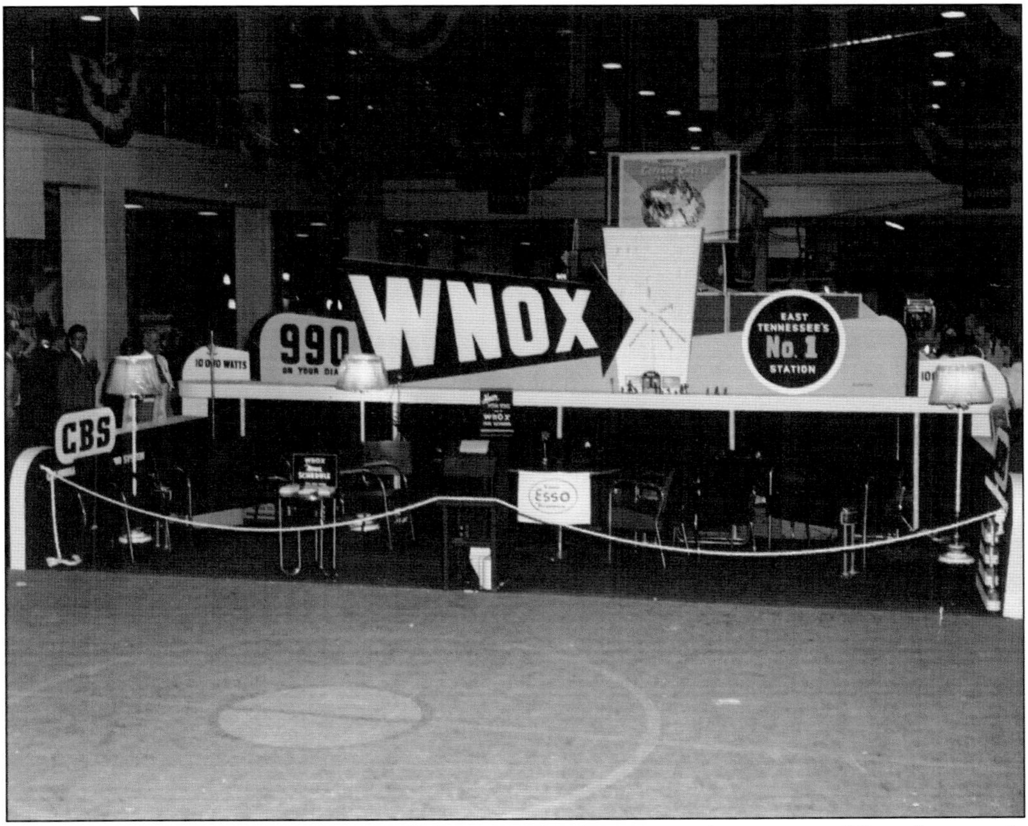

WNOX was a major player in Knoxville business development and promotion. This photograph is of a station display set up at a local event. The WNOX sign is pointing to an art-deco, three-dimensional mockup of the station's studios on Gay Street.

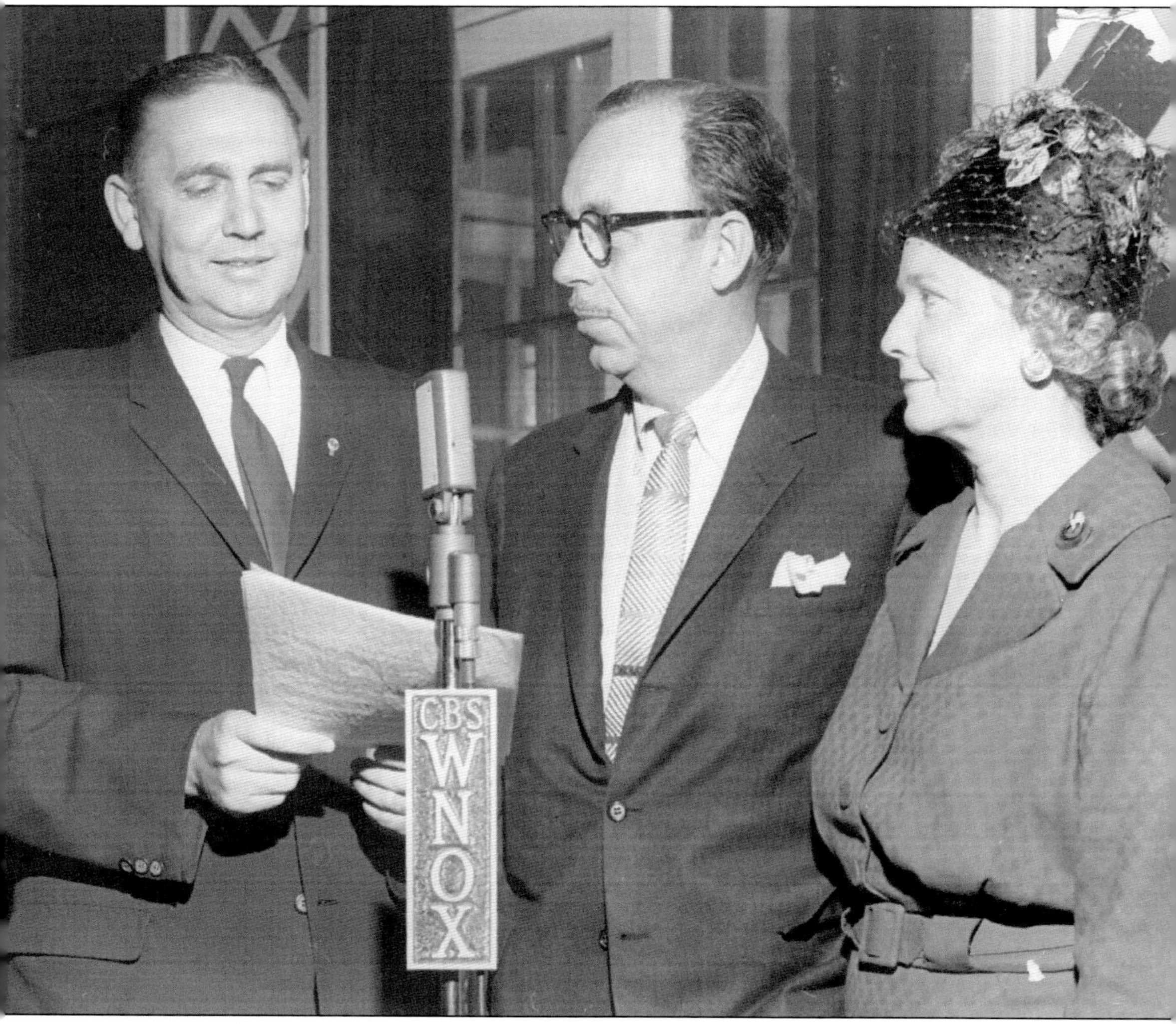

WNOX was honored many times for its service to the community. Pictured here is then-mayor and future Knoxville congressman John Duncan (on far left) presenting a proclamation to WNOX general manager Dick Westergaard and his wife, Ruby.

Four

The Mid-day Merry-Go-Round and Tennessee Barn Dance

These are the programs most talked about, written about, and enshrined in the annals of country music. For many, they are the trademark shows of WNOX that remain to them the station's single greatest achievements. They were the brainchild of announcer and program director Lowell Blanchard. The *Mid-Day Merry-Go-Round* was a variety show of local talent that featured a little of everything. It had its own orchestra that played big band, soloists who performed religious songs, and even a song or two from the classical repertoire. The *Tennessee Barn Dance* was exclusively dedicated to the new sound Lowell Blanchard was discovering and promoting.

The *Mid-Day Merry-Go-Round* began at 12:30 and ran until 1:45 six days a week. With the last band out on Saturday, they broke it down and set up for the *Barn Dance.* Blanchard invited the listeners in to see the programs that featured what then could only be called "hillbilly bands." His purpose in doing so was in keeping with management's mantra to entrench the station with East Tennessee listeners. In many cases, they were little-known family groups that played in local churches, at picnics, or family gatherings. Blanchard culled them personally. He would hear rumors of a good guitar, banjo, or fiddle player and drive into the countryside to find them.

Most would at first rebuff the slick talker from the city and refuse to play on stage. He had to work them, convince them that if their church, families, and communities liked them, they owed it to them to come in and just try to play a song or two for the radio so everyone could hear. He would assist them in devising a stage presence, be their cheerleader and mentor when it came to the microphone. Blanchard was the consummate showman and became a master of stagecraft. He would remember the little details of where he found the musicians and how he heard about them, and it would be inserted into his stage patter, which made him one of the great masters of ceremonies.

Blanchard was educating himself as well. Those programs were his inventions, and he wrote his own daily scripts to present on stage. He researched and paid attention to his audiences so he could develop a sense of timing in his delivery. The straight man from Illinois with serious ambitions developed a "country" sense of humor, learned the accent, and, when he hit the stage, was ready for business. He did not care if 10 or 10,000 were in the seats in front of him. His job was to announce and put on a show—not worry about audience numbers or the lack of them. He sold the station by selling the listeners on the next act in queue. It was a split audience. On one hand, there were the listeners to WNOX, and on the other hand, there were those in the seats in front of him that also became part of the broadcast program. In the transition between acts, he would tell jokes, sell a product, and fill the time necessary before turning it over to the next performer.

From the top floor of the Andrew Johnson Building in 1936, the programs debuted. The musicians and the crowds started as a trickle and gradually increased. Westergaard rented a boxing arena where the musicians would stand in the ring and play, but the station soon outgrew it. He then rented space at the Knoxville Market House, and it outgrew that venue as well. Westergaard decide to charge 5¢ admission, and people were turned away; he raised it 15¢ and all the way to a quarter and people kept coming. A turn away from the programs was one less friend to the station, and standing room only was the rule of the day for WNOX.

The move to 110 South Gay Street gave the station its very own auditorium, and there the program really took root. The audience numbers and the workday lunch crowd would descend en masse on the station to see the programs.

"I train them and Nashville gets them," was the quote Lowell Blanchard said one day when asked about the string of country music superstars he discovered and paraded across the stages of WNOX. He is the single reason the question is asked why Knoxville did not become the "Home of Country Music" and was comfortable with being its "cradle" instead. There were business reasons. WNOX was a hugely successful station. Scripps Howard did not want to mess with the success, and Nashville's WSM pulled an AM signal online with a 100,000-watt transmitter and a signal that could be heard in more places at night in the United States than WNOX's.

In 1954, the station bought the Whittle Springs Resort and Hotel in Fountain City. By then, the country music sound birthed at WNOX was sitting on record labels across the nation. The acts and the audiences kept coming, and the new facility—built to house both a television and radio station—was the only live studio of its kind in the world. It was, to many people, the centerpiece of Knoxville travel and tourism and a shrine to both country music and its most beloved promoter.

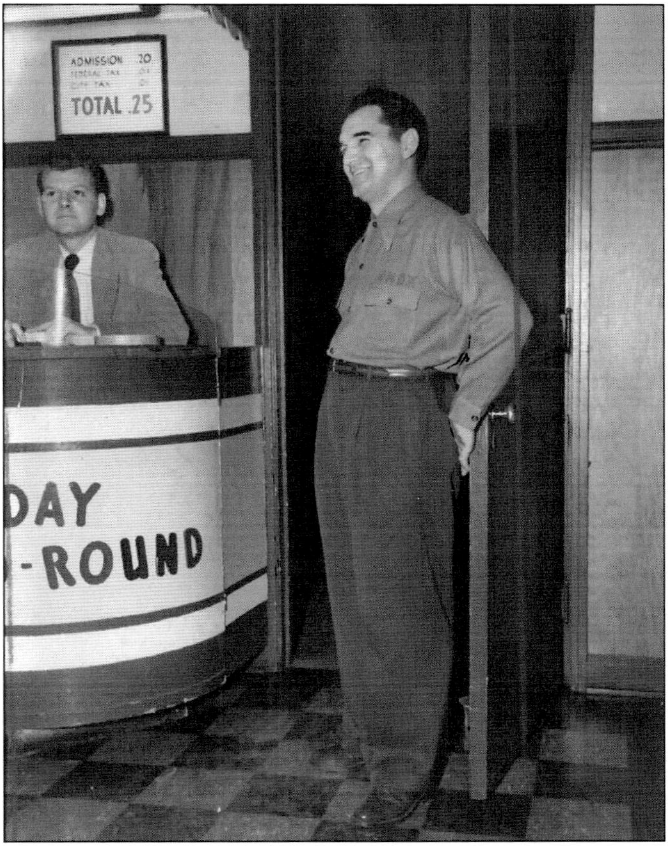

At right is a photograph of the entrance to the *Mid-Day Merry-Go-Round*. From left to right are Clinton Danner, who sold the tickets, and Dennis Shehan, who collected them. The ticket price at the time this photograph was taken was already 25¢, including 5¢ tax. Below is Shehan's wife, Louise, who sold popcorn to the audience in attendance. It was common to see downtown workers in the auditorium eating lunch while watching the program. (Right, courtesy of E. W. Scripps Company.)

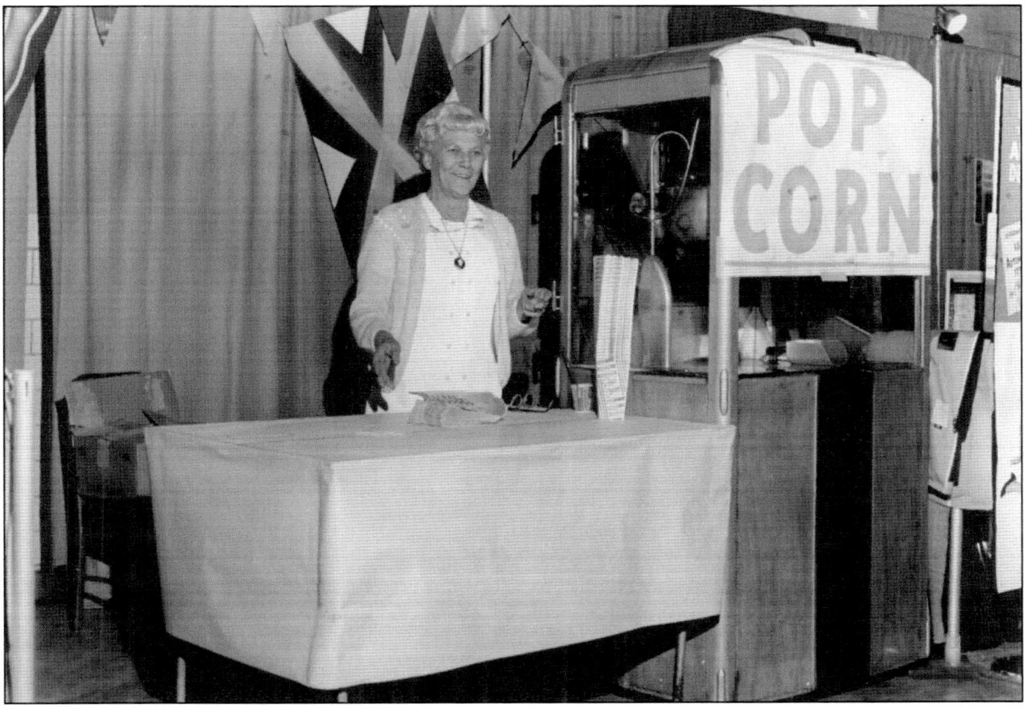

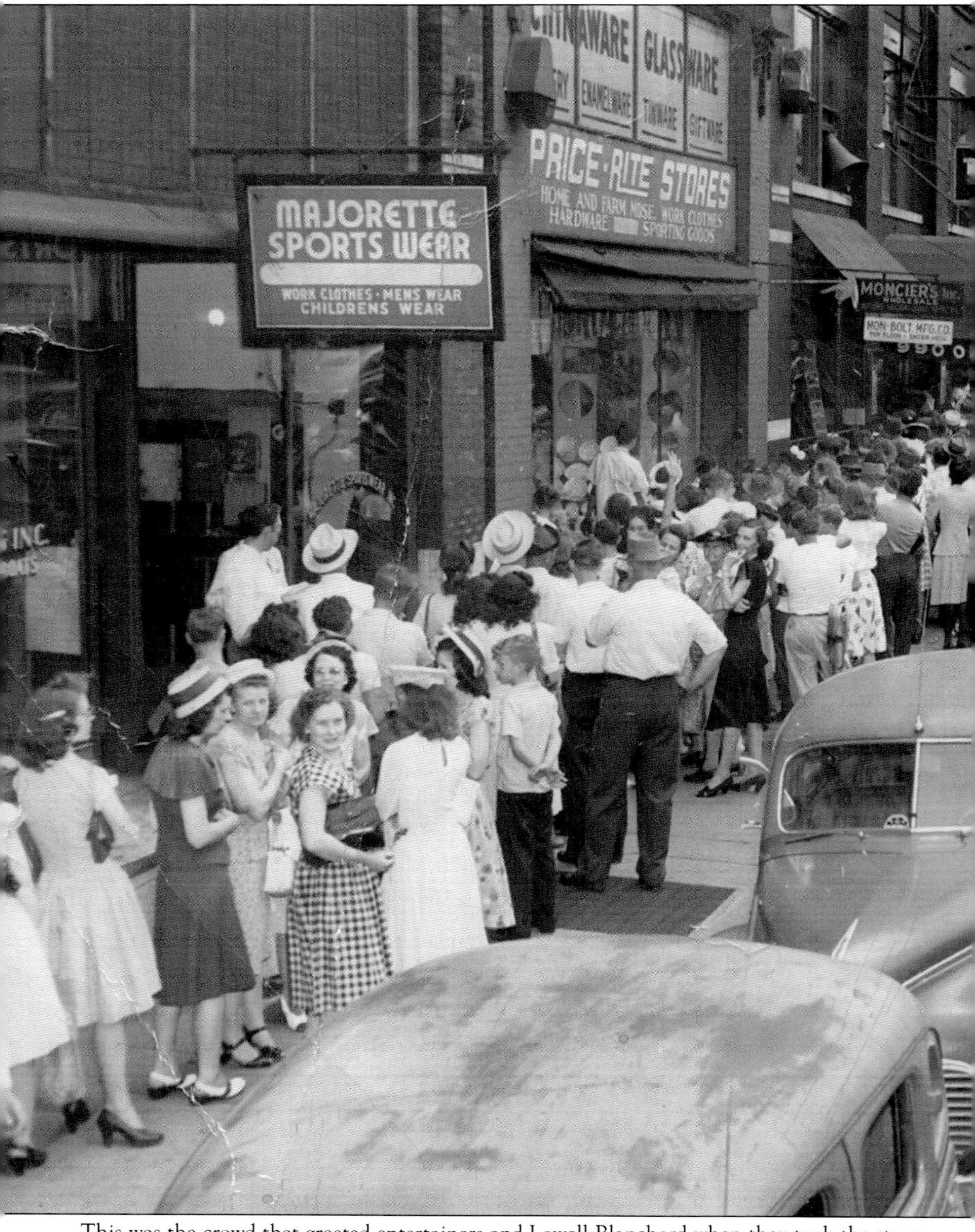

This was the crowd that greeted entertainers and Lowell Blanchard when they took the stage. The majority of audience members were from Knoxville and East Tennessee, but the program's appeal attracted people from across the nation to WNOX's Gay Street Auditorium. Policemen

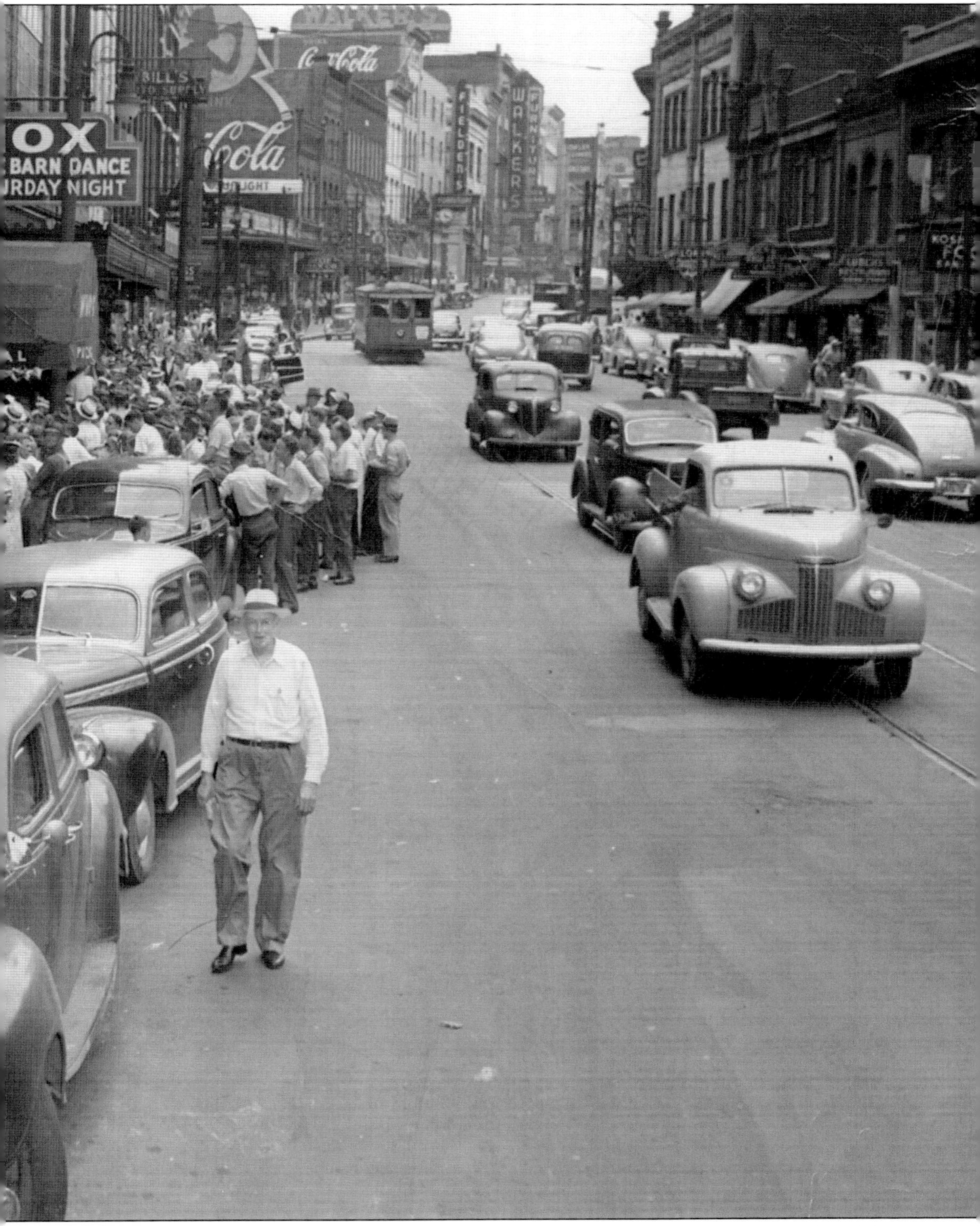

were dispatched daily to handle the massive crowds and move them as quickly as possible off the sidewalk. It sometimes frustrated musicians when they tried to off-load instruments and equipment and police would not give them time to get unloaded before ordering the cars to move.

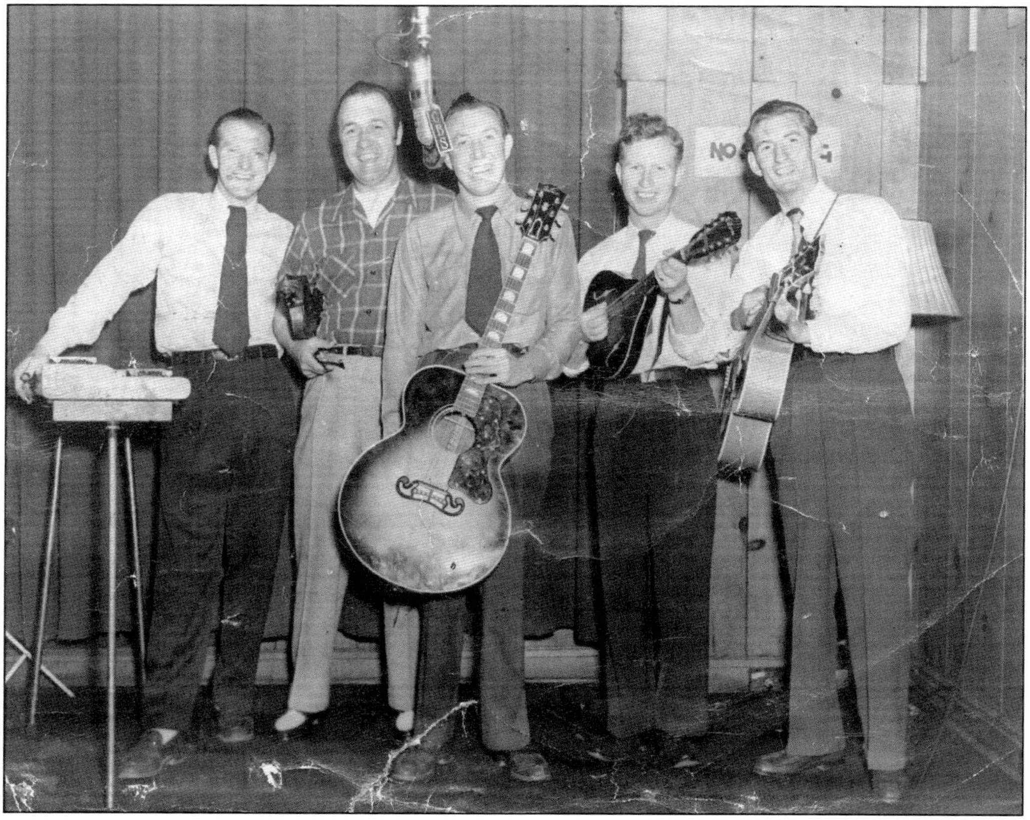

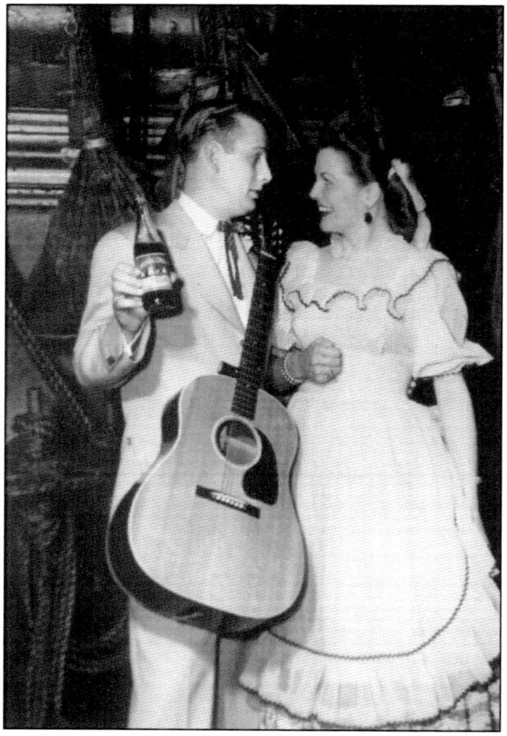

Carl Story and the Rambling Mountaineers was a popular band on the *Mid-Day Merry-Go-Round*. Carl Story went on to become known as the father of bluegrass gospel music and recorded albums for different labels through the 1970s. He was a fixture on the bluegrass festival circuit, and his band was known as a training ground for many future country artists.

Martha Carson was a famous gospel singer who rose to fame and popularity on the *Mid-Day Merry-Go-Round*. Next to her is singer George Morgan, who made the song "Candy Kisses" a popular song. Here the two entertainers are doing a live commercial for Royal Crown Cola.

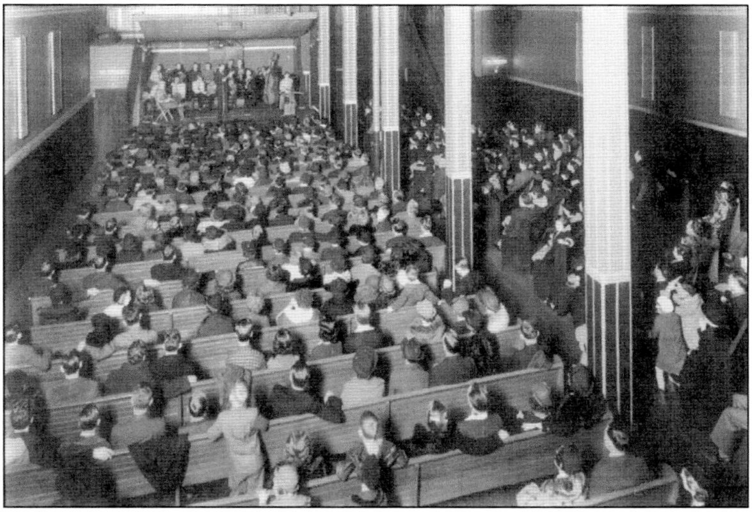

This photograph was taken of a *Mid-Day Merry-Go-Round* performance at the Gay Street Auditorium. The pews accommodated many more people than the normal auditorium seating. On stage is announcer Lowell Blanchard with the WNOX house band, where many station employees performed in addition to their regular jobs.

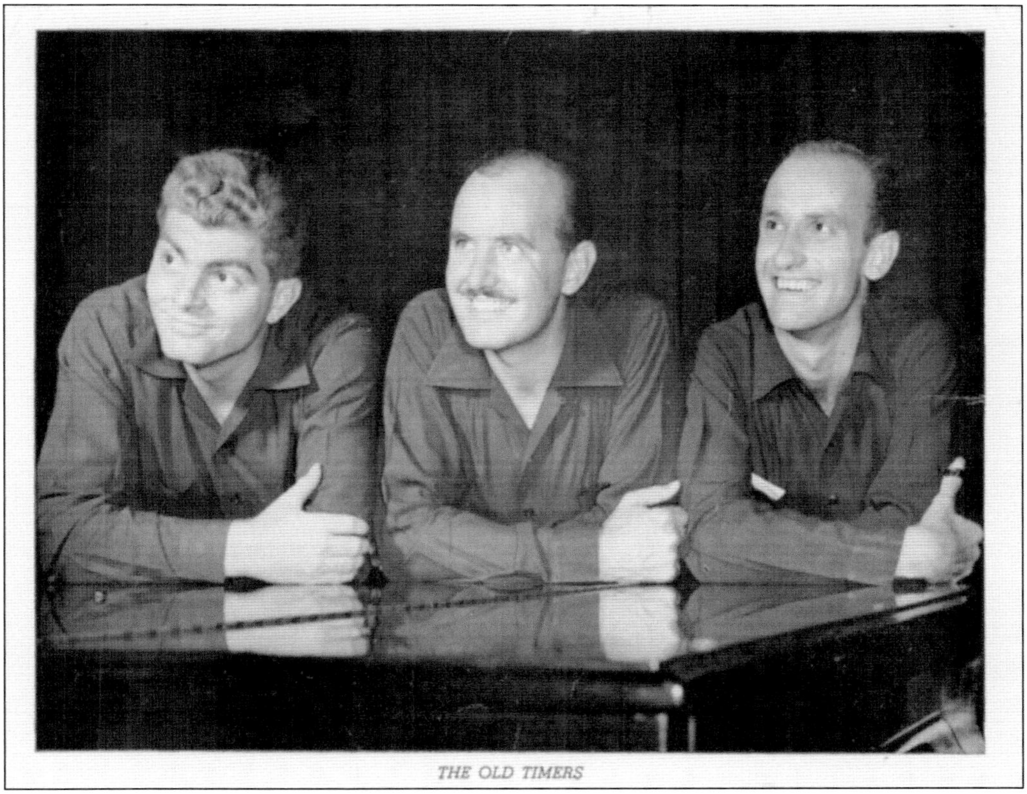

From left to right are Red Kirk, Archie Campbell, and Speedy Krise, Knoxville radio performers who performed for WNOX and other stations in the city. Campbell created the character "Grandpappy," which is one of the most remembered comedy routines on the *Mid-Day Merry-Go-Round*. It a role he reprised throughout his career. He was also an avid golfer, and musicians often complained about having to track him down on a golf course, which sometimes made them late to performances.

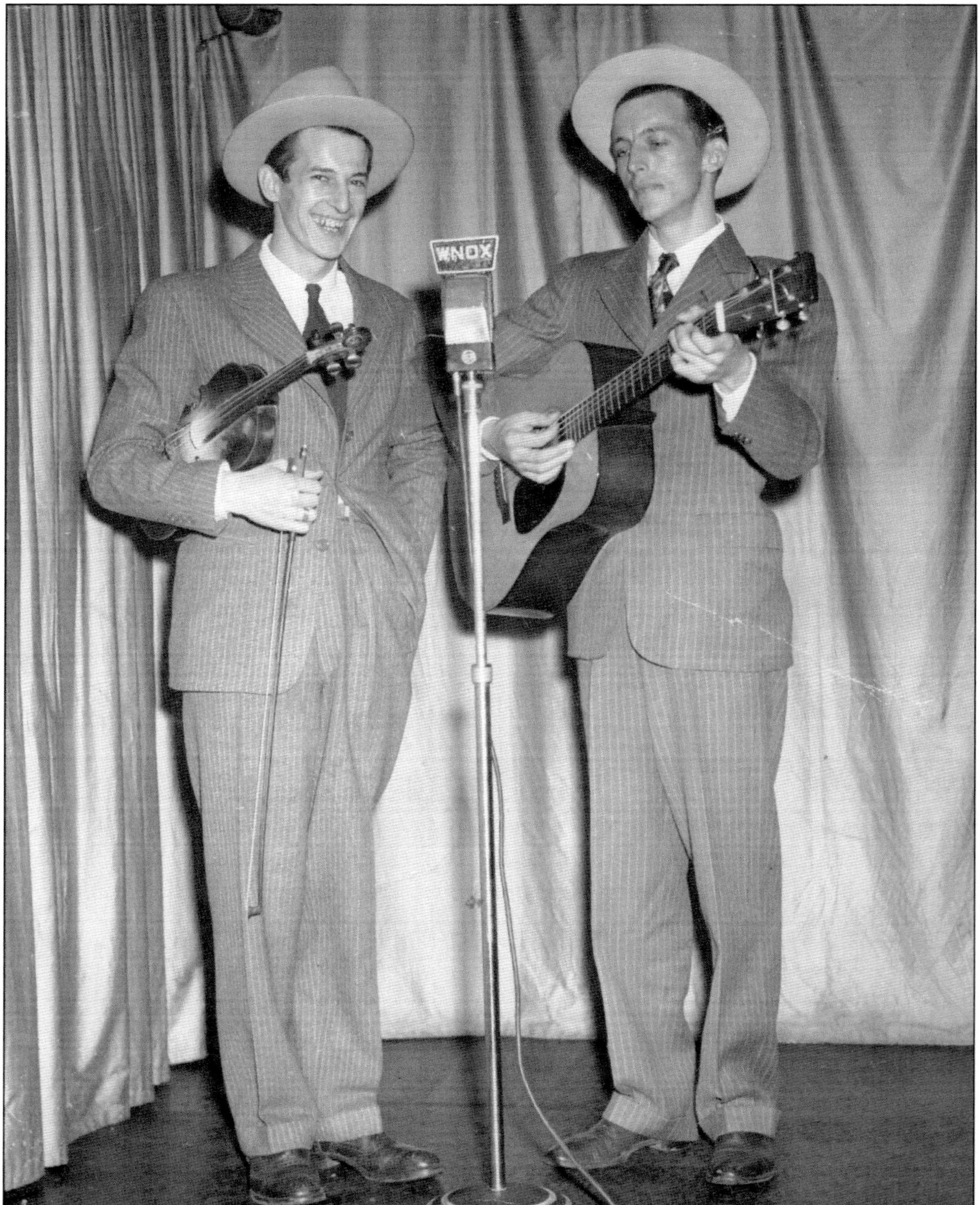

From left to right are Joe and Bill Burchfield, who were regulars on WNOX's *Mid-Day Merry-Go-Round* and the *Tennessee Barn Dance* in the late 1940s. They performed primarily old-time bluegrass and gospel music. The family was a close associate of Archie Campbell, and they remained friends throughout their lives. (Courtesy of Patsy Robertson.)

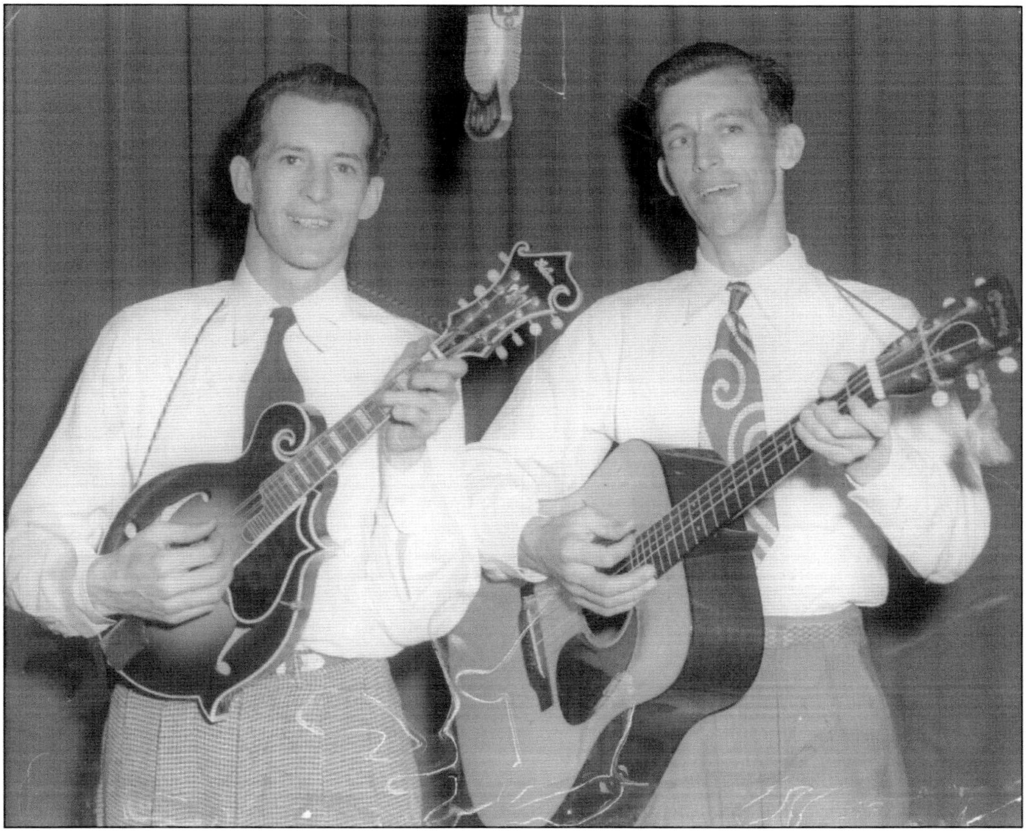

The Burchfields, like many families in East Tennessee, were musically oriented. Before radio and television, music was a family's only entertainment. Musical instruments, singing, and performing for each other were how they would spend weekends. The Burchfields earned the respect of music producers with their act and, in 1947, were signed to a Capitol Records recording contract. The East Tennessee sound they brought to the stage would be an influence on other recording stars as country music began getting a foothold in American culture. (Both courtesy of Patsy Robertson.)

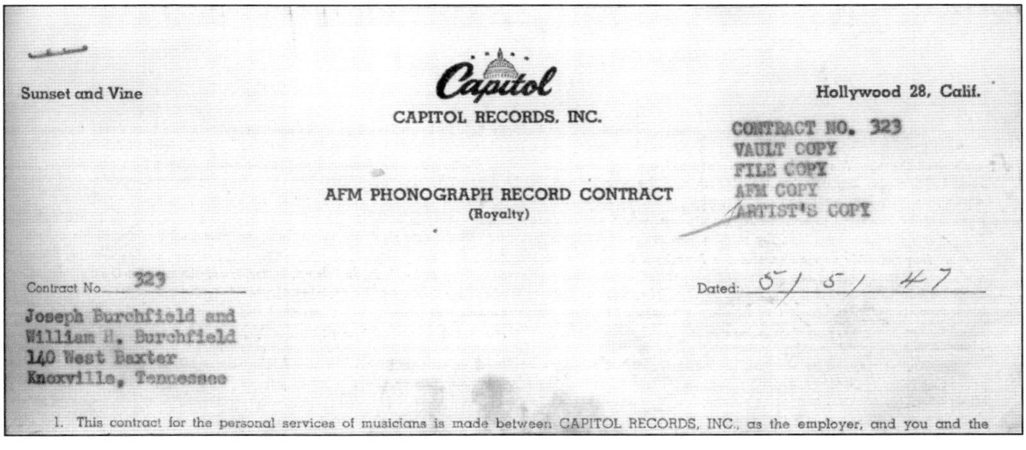

Sunset and Vine

Capitol

CAPITOL RECORDS, INC.

Hollywood 28, Calif.

CONTRACT NO. 323
VAULT COPY
FILE COPY
AFM COPY
ARTIST'S COPY

AFM PHONOGRAPH RECORD CONTRACT
(Royalty)

Contract No. 323

Dated: 5/5/47

Joseph Burchfield and
William H. Burchfield
140 West Baxter
Knoxville, Tennessee

1. This contract for the personal services of musicians is made between CAPITOL RECORDS, INC., as the employer, and you and the

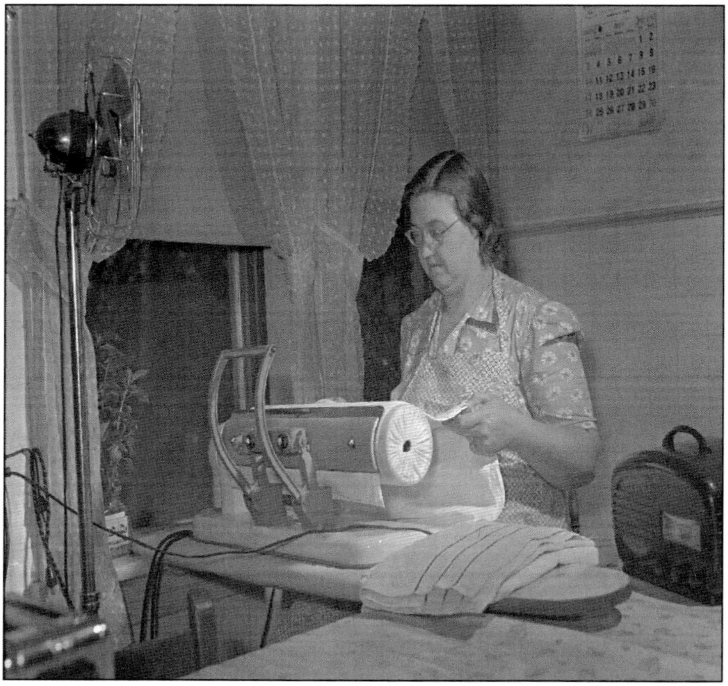

This photograph taken of a Knox County homemaker for the Tennessee Valley Authority shows her pressing clothes with all the electrical appliances that made life easier, including a radio on the far right. As radio became popular and models got smaller and more affordable, it was not unusual to see them in barns and workplaces in East Tennessee. It was the car radio, however, that put the electronic medium everywhere and made it America's most powerful advertising tool.

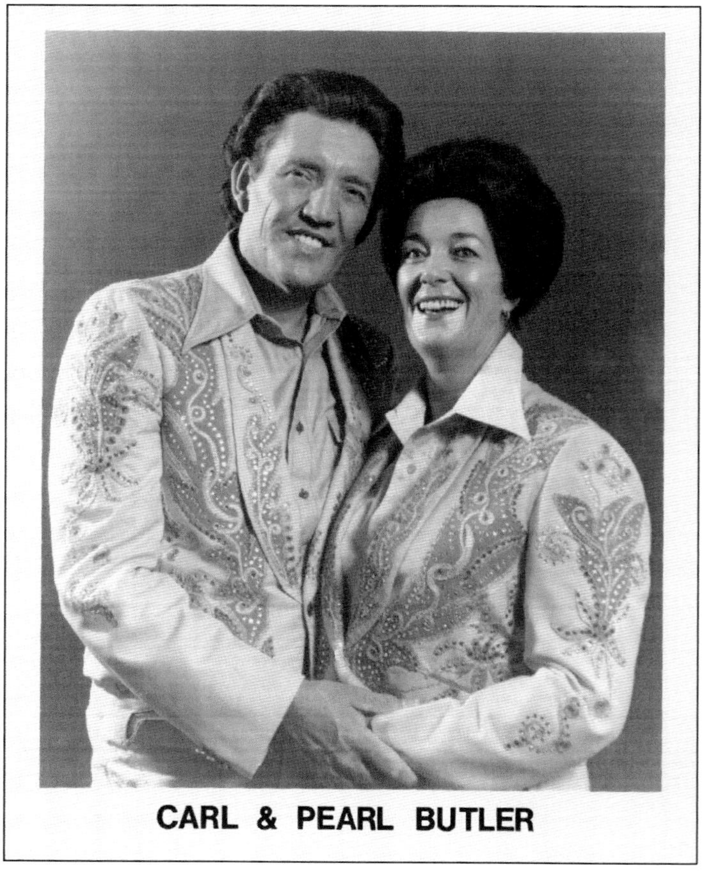

Another popular recording duo was Carl and Pearl Butler. The couple met in Knoxville when he was performing on WNOX. They became a national sensation when they recorded "Don't Let Me Cross Over" and spent almost three months at the top of the Billboard Charts in 1962. They went on to a successful Nashville recording career and became fixtures in the country music industry. (Courtesy of John Hitch.)

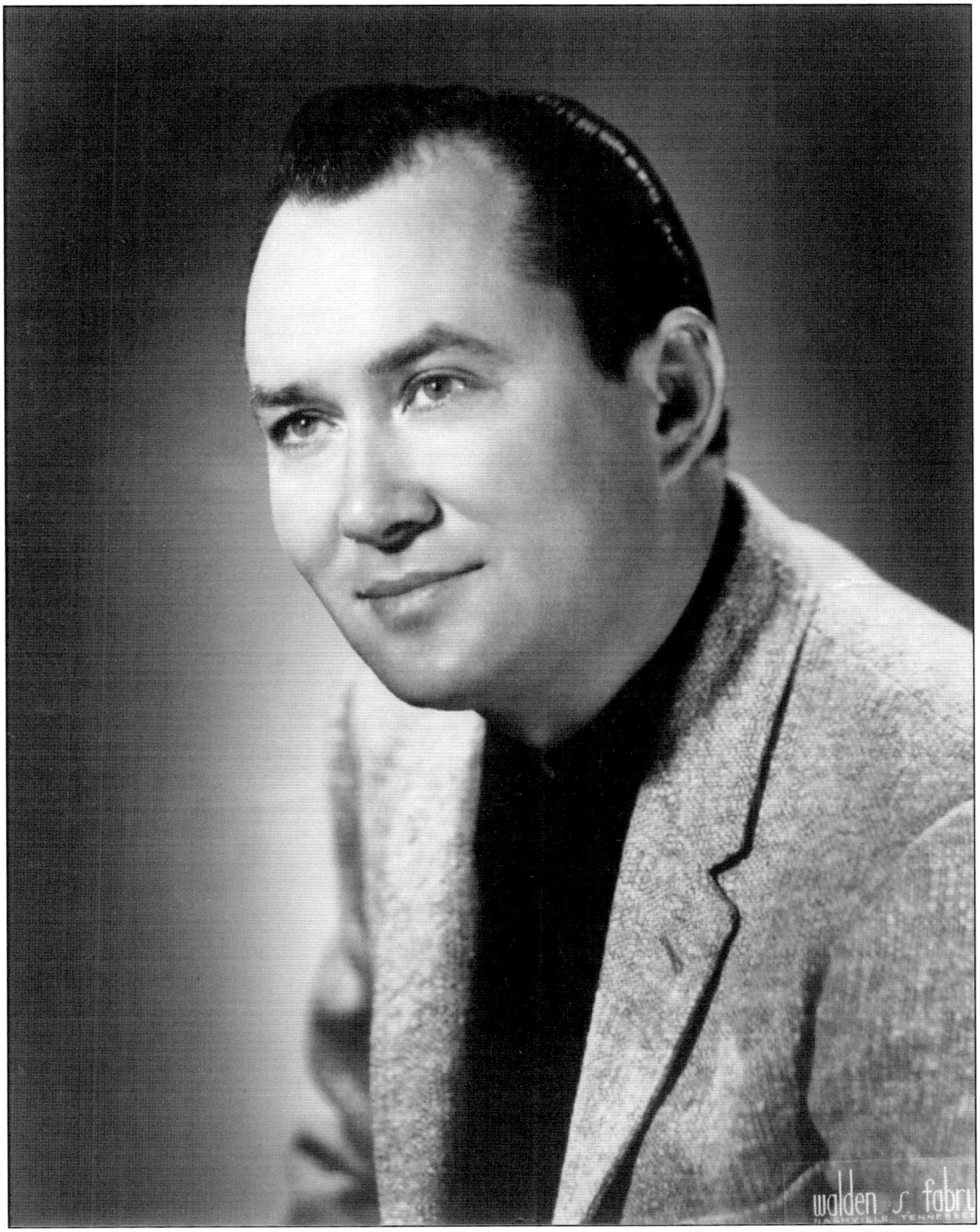

Country Music Hall of Famer and WNOX standout Don Gibson resided in Knoxville and collaborated with songwriter Arthur Q. Smith. He started his career at WNOX and recorded such classics as "I Can't Stop Loving You" and "Oh Lonesome Me." It was a double-sided, number one single for RCA on both the country and pop charts of 1957. He was known as a songwriter and would spend days in Knoxville collaborating with songwriter Jim Pritchett (Arthur Q. Smith). Gibson had a string of country music hits that stretched from 1957 to well in the 1970s. (Courtesy of Ron Anderson.)

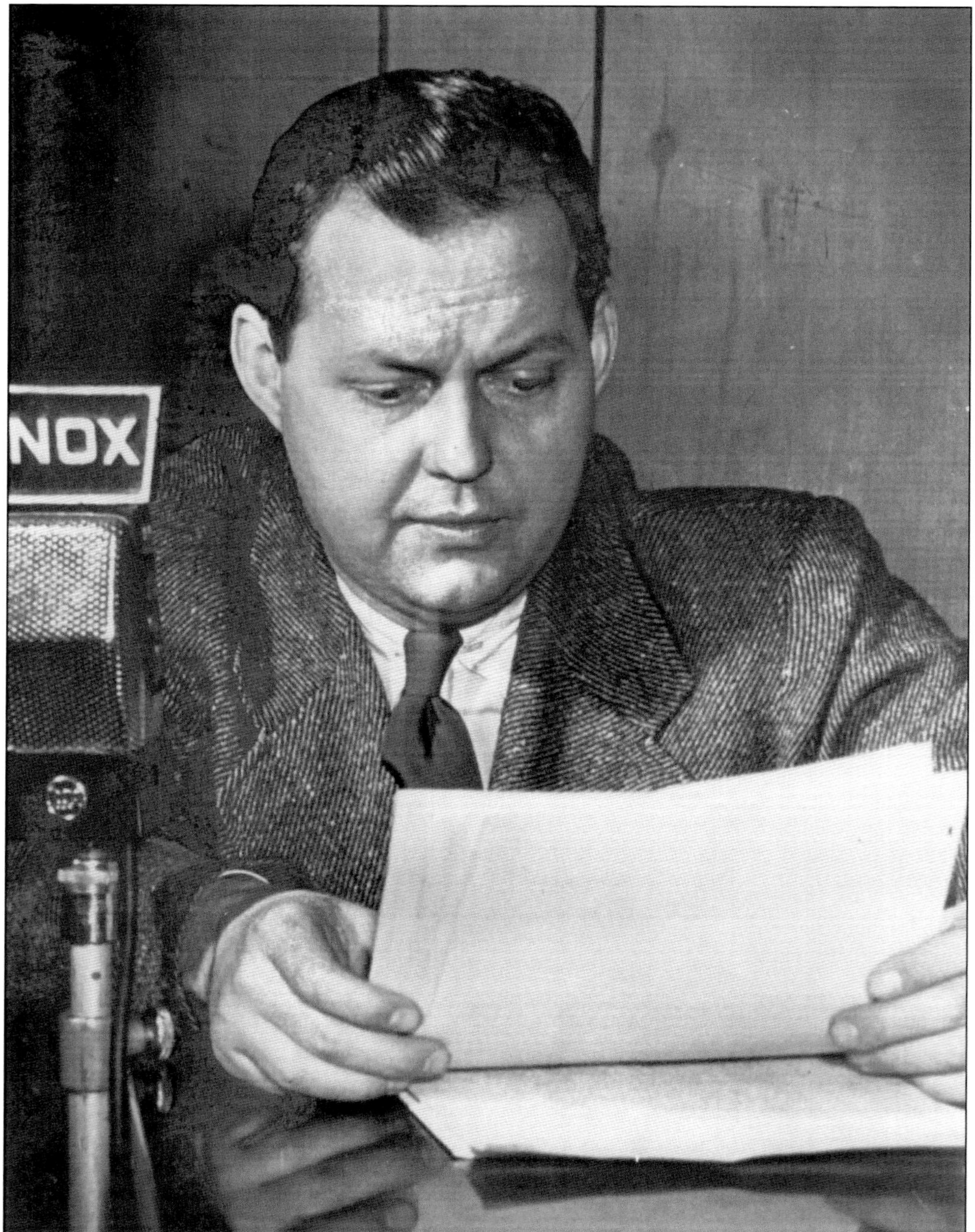

Lowell Blanchard is photographed here behind the WNOX microphone. He scripted prior to the programs the humor, wit, and on-stage abilities that came so easily and were so admired by fans. Few people knew how much research and creative writing when into the *Mid-Day Merry-Go-Round* and *Tennessee Barn Dance*. (Courtesy of Harley Sweet.)

Another famous Carl to perform on WNOX's *Tennessee Barn Dance* and *Mid-Day Merry-Go-Round* was Carl Smith. The Maynardville, Tennessee, native grew up listening to WNOX. He was one of the biggest country music hit makers in the 1950s and 1960s and worked with Hank Williams, Ernest Tubb, and other greats in the industry. Every album he made in the early 1950s made the Top 20, and four of his releases were "triple crown winners" in the industry—reaching number one on all three of Billboard's country music charts. (Courtesy of John Hitch.)

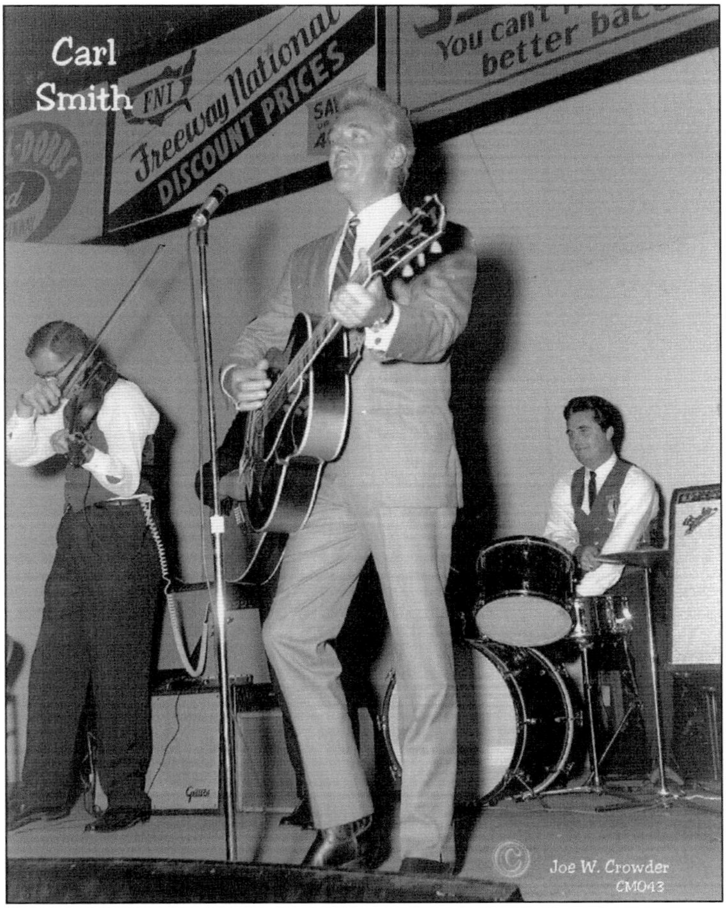

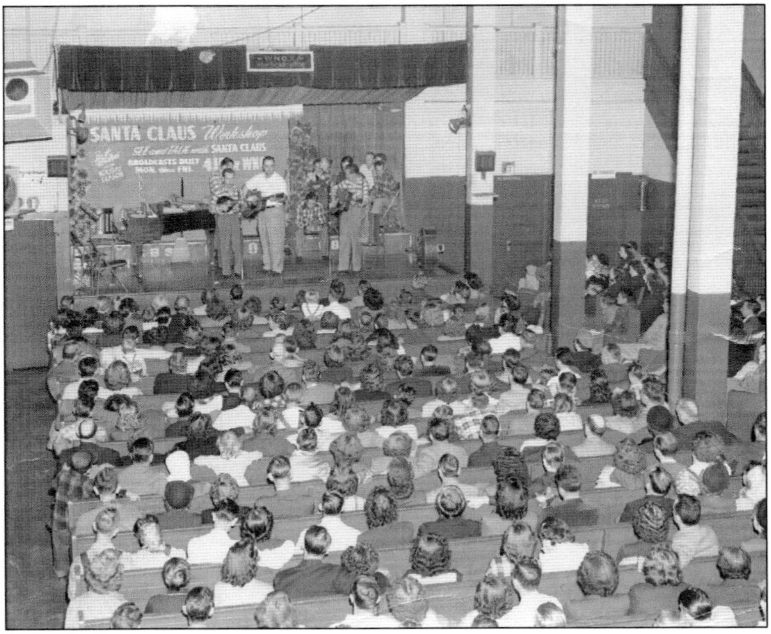

This photograph was taken of the *Mid-Day Merry-Go-Round* during the holiday season when the radio station held its annual Santa's Workshop, where children could come and meet Santa Claus. It was a holiday ritual for the station, as was its participation in the Empty Stocking Fund, which was coordinated through the *Knoxville News Sentinel*.

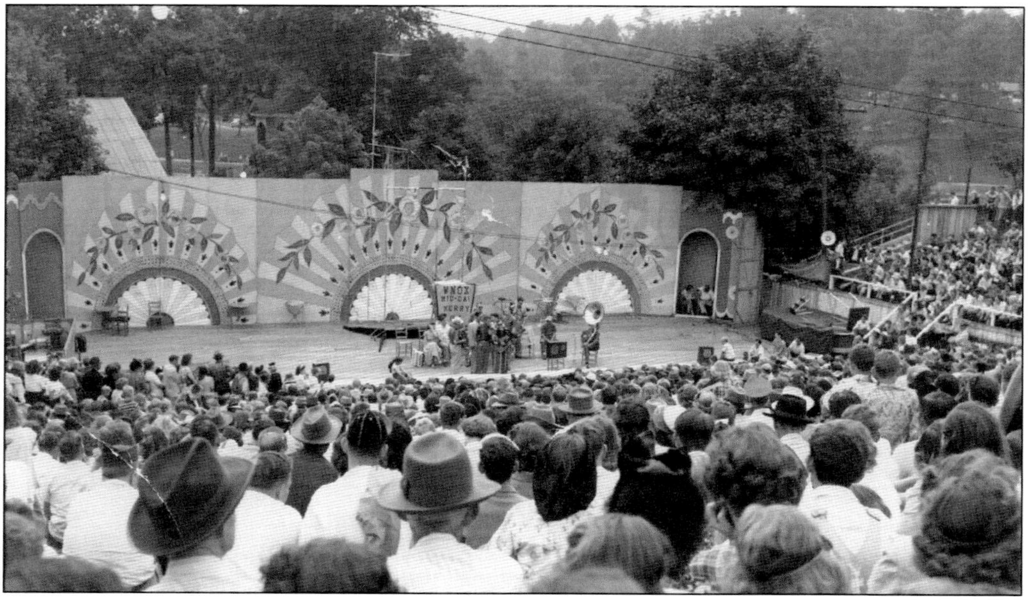

Another festive time in East Tennessee was the annual Tennessee Valley Agricultural and Industrial Fair held at Chilhowee Park in east Knoxville. This is an image of Earl Scruggs (third from the left) and Lester Flatt (fifth from the left) performing for the crowd. They got their start on the radio station and went on to become the first bluegrass artists to perform at New York City's Carnegie Hall in 1962. The fair is one of the city's biggest annual events and attracts entertainers from across the nation. (Courtesy of Harley Sweet.)

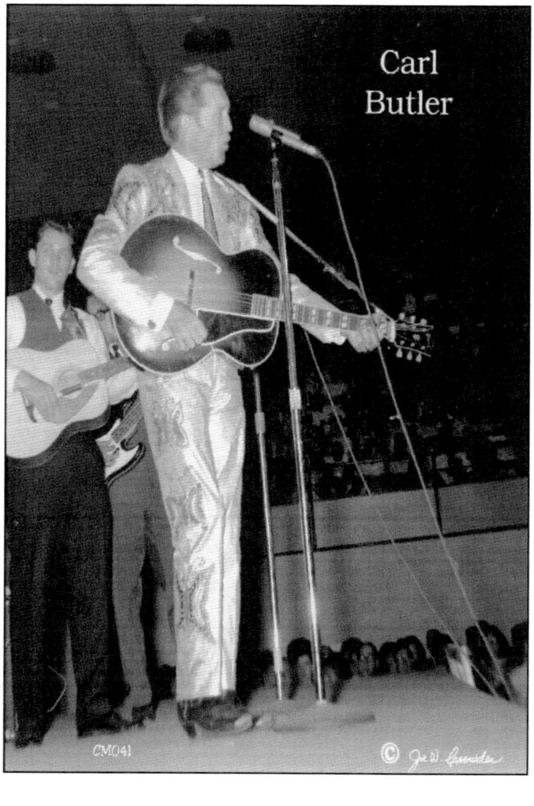

Pictured here is Carl Butler performing as a solo act at WNOX's *Tennessee Barn Dance*. While Butler was a good songwriter, the songs "I Wouldn't Change You If I Could," "If Teardrops Were Pennies," and "Crying My Heart Out Over You" were actually written by Arthur Q. Smith. Butler purchased the latter two, which became country music classics, from Smith and put his name on them. (Courtesy of John Hitch.)

The Carter Family played the *Mid-Day Merry-Go-Round* on a number of occasions. They were one of the most popular bands of their day. On the right, from left to right are June Carter, Mother Maybelle Carter, Anita Carter, and Helen Carter. Their recordings of folk songs like "Wildwood Flower" earned them a loyal following and a string of recording hits. Below, Helen (left), Anita (center), and June Carter get ready to perform on the *Mid-Day Merry-Go-Round*. (Both courtesy of Roy Garrett.)

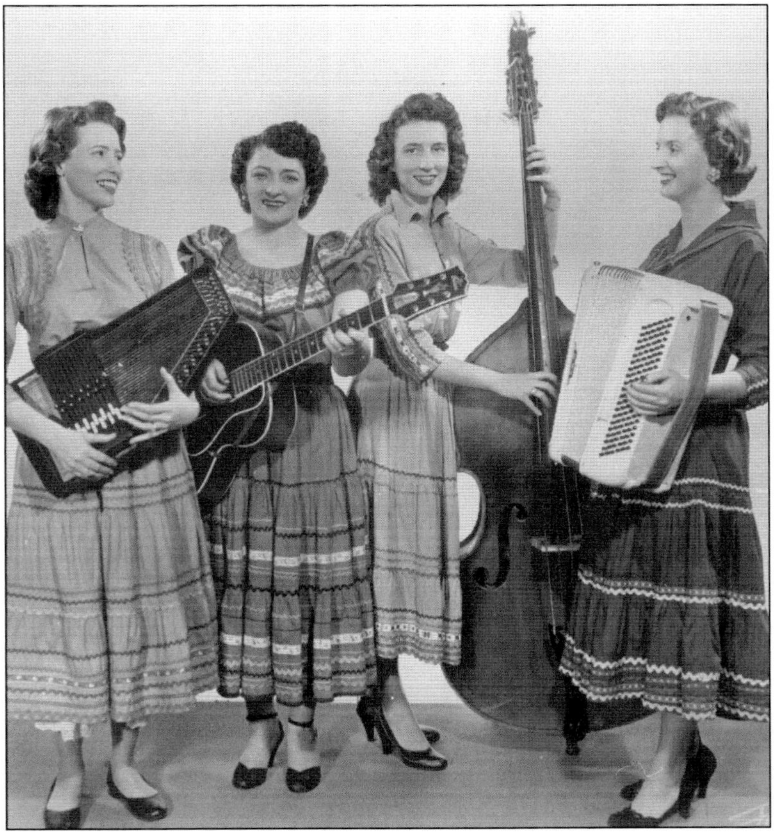

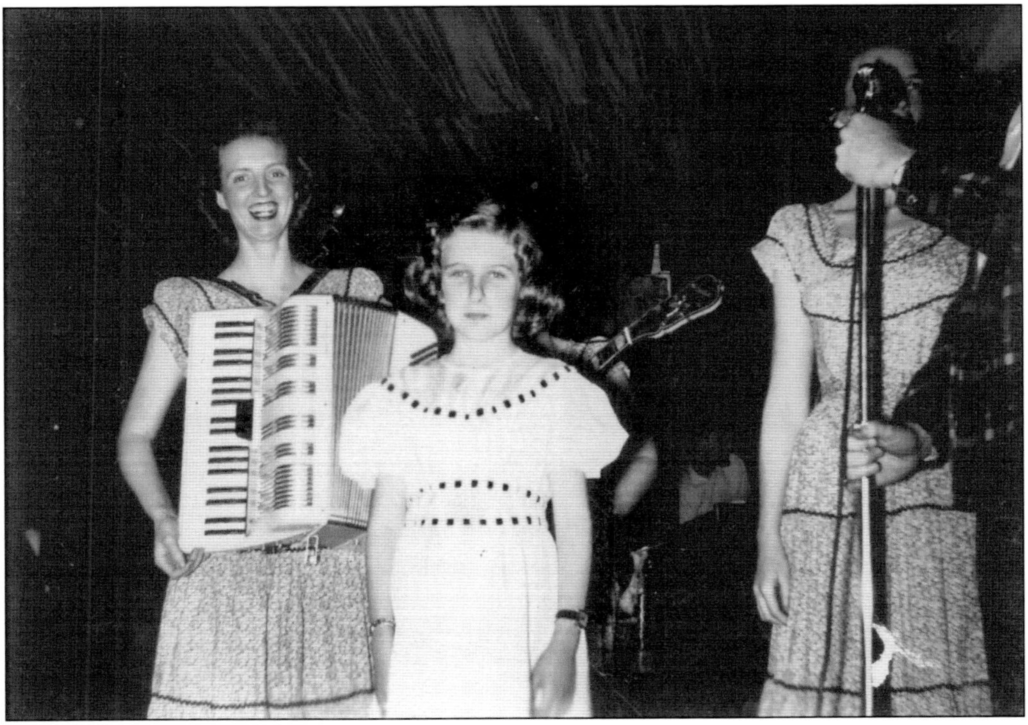

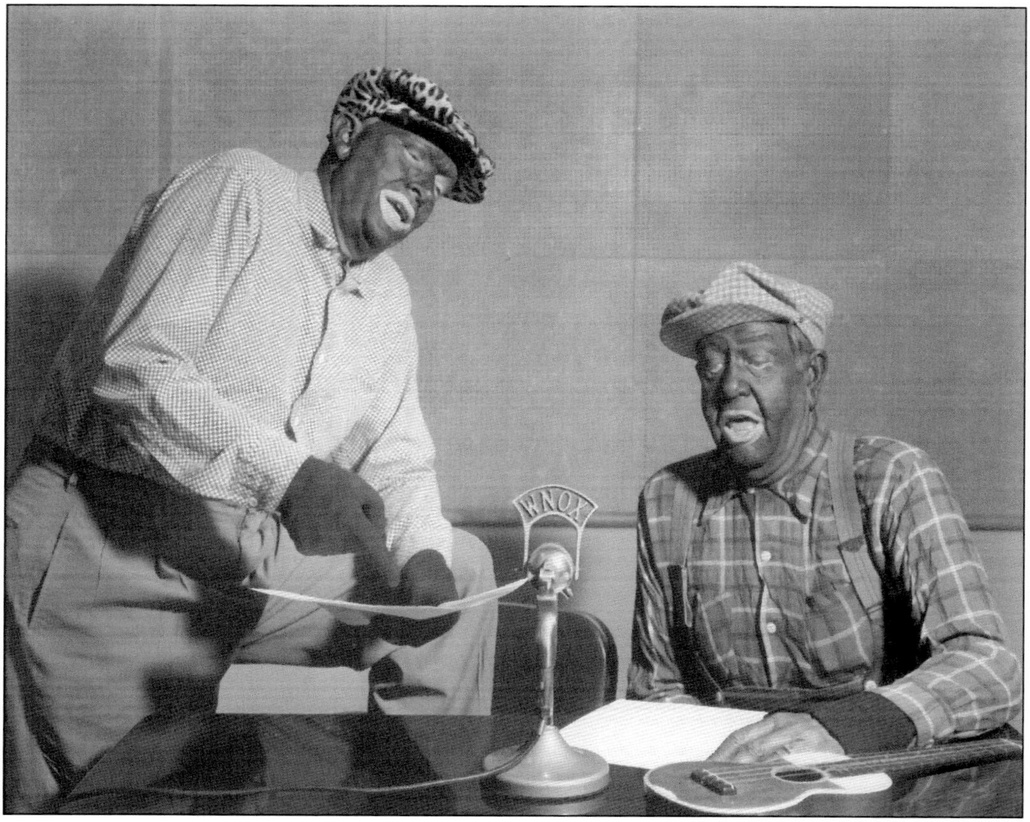

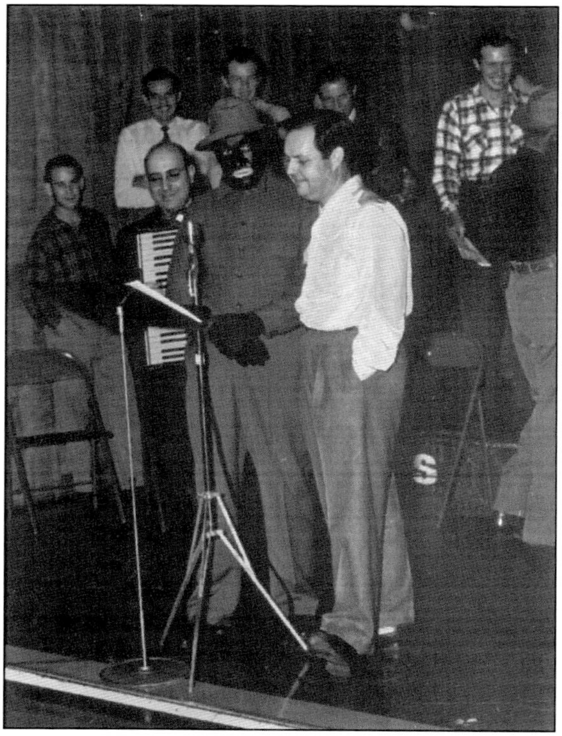

Jam Up and Honey were the Southern answer to Amos and Andy. Lee David Wilds (Honey), on the right, was a vaudeville blackface comedian who took his act to the Grand Ole Opry in 1932 with Lee Roy "Lasses" White. After White left for film roles in Hollywood, Wilds formed Jam Up and Honey. His partner, Jam Up, was a character played first by Tom Woods and then Bunny Biggs. Wilds was the creator of the Opry tent shows, which featured a variety of popular acts like Eddie Arnold, Ernest Tubb, Hank Williams, Roy Acuff, and Bill Monroe and toured across the nation in the 1940s. According to legend, he was the one who gave the nickname "Bocephus" to Hank Williams Jr. On the left, Wilds is doing a comedy routine on WNOX's *Mid-Day Merry-Go-Round*. From left to right are Tony Musco, Honey Wilds, and Lowell Blanchard. (Above, courtesy of Harley Sweet; left, courtesy of Roy Garrett.)

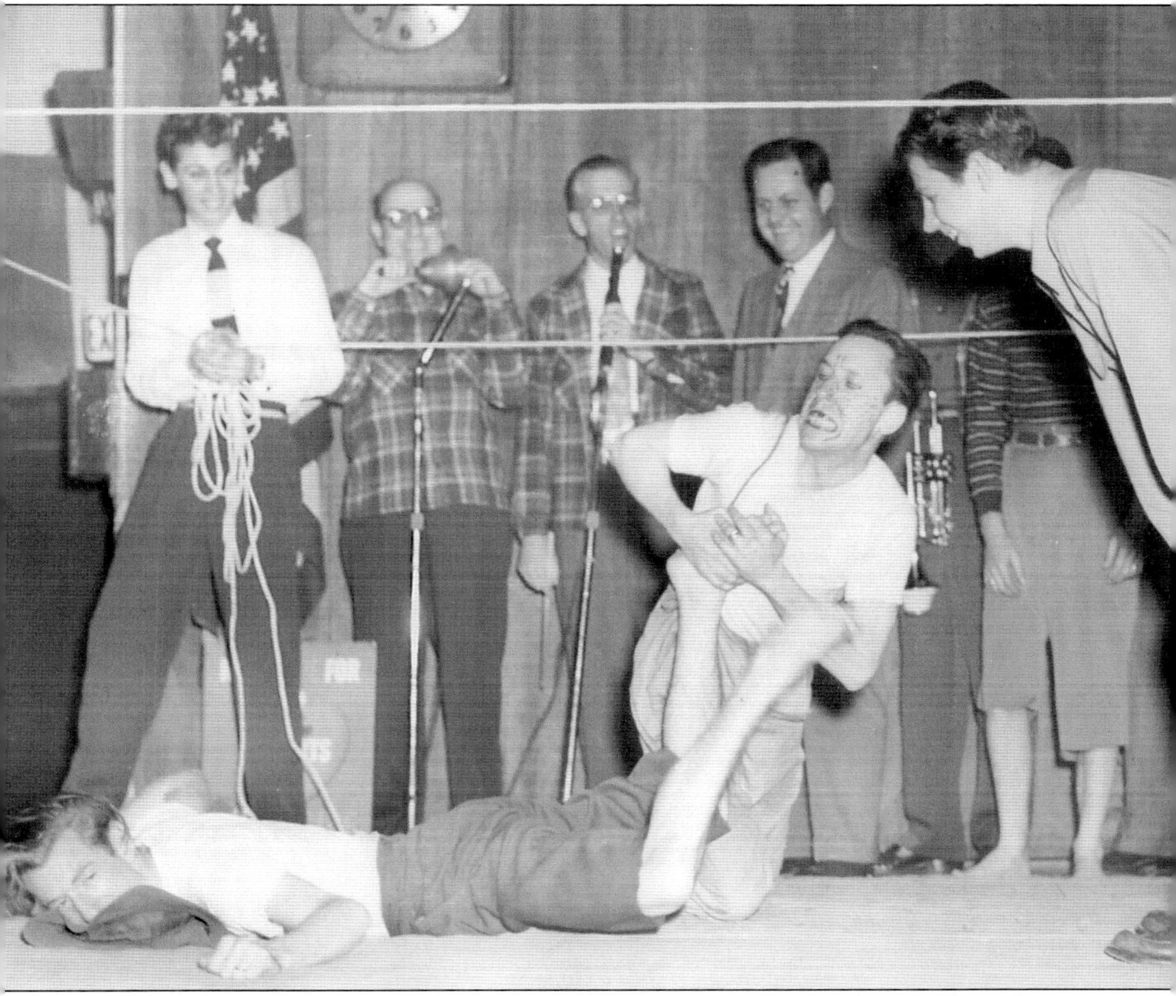

This is a promotional photograph of a comedy sketch with "Hot Shot Elmer" done for the *Mid-Day Merry-Go-Round* in 1952. The comedic character created by Bill Carlisle (lying on the mat) would put on boxing gloves and pretend to fight with cast members at the *Mid-Day Merry-Go-Round* and *Tennessee Barn Dance*. It was one of the most hilarious and loved parts of the programs. Bill Carlisle started his career with his brother in the mid-1930s and rose to national prominence on the WNOX stage. (Courtesy of Harley Sweet.)

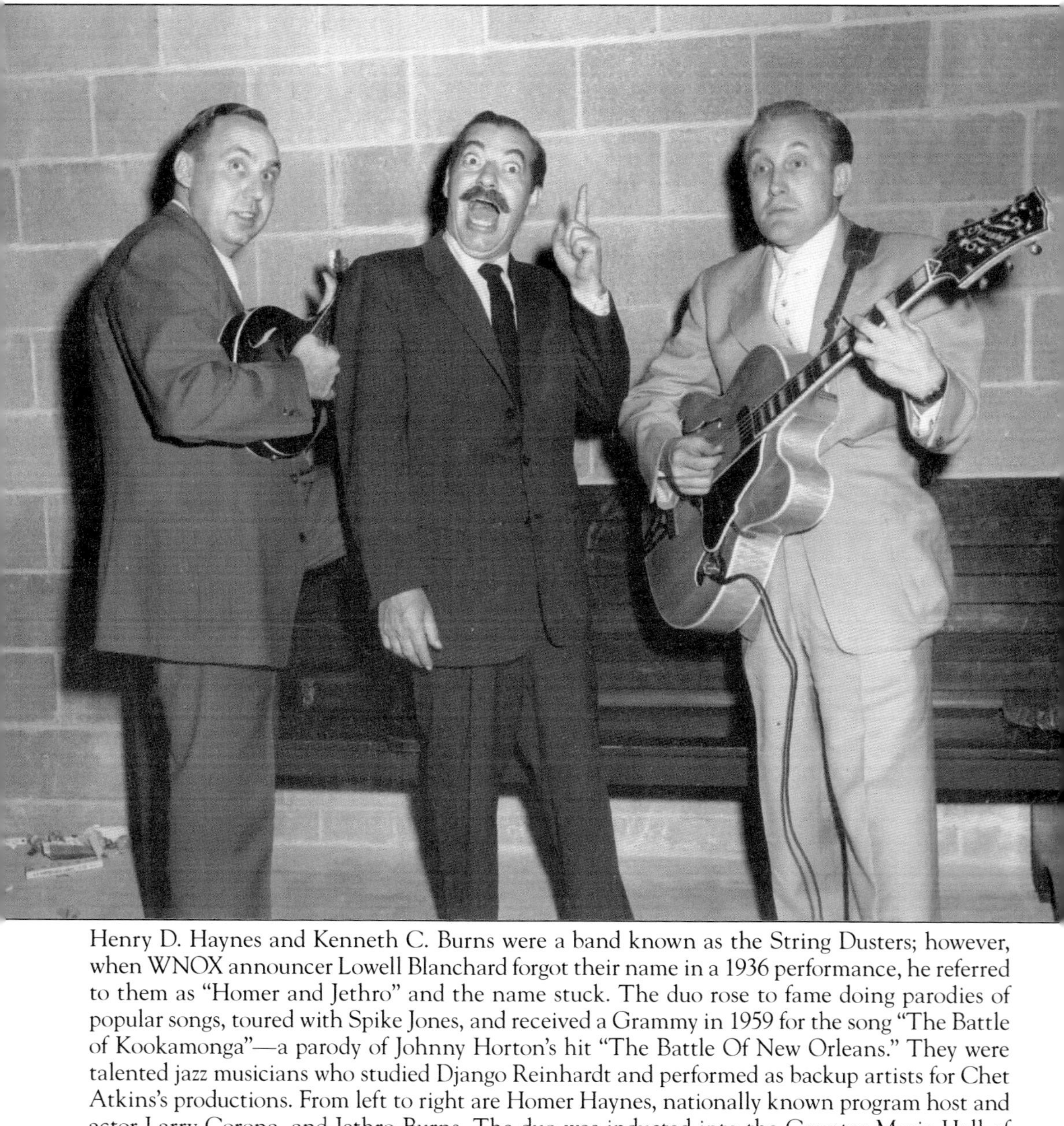

Henry D. Haynes and Kenneth C. Burns were a band known as the String Dusters; however, when WNOX announcer Lowell Blanchard forgot their name in a 1936 performance, he referred to them as "Homer and Jethro" and the name stuck. The duo rose to fame doing parodies of popular songs, toured with Spike Jones, and received a Grammy in 1959 for the song "The Battle of Kookamonga"—a parody of Johnny Horton's hit "The Battle Of New Orleans." They were talented jazz musicians who studied Django Reinhardt and performed as backup artists for Chet Atkins's productions. From left to right are Homer Haynes, nationally known program host and actor Larry Corona, and Jethro Burns. The duo was inducted into the Country Music Hall of Fame in 2001. (Courtesy of Harley Sweet.)

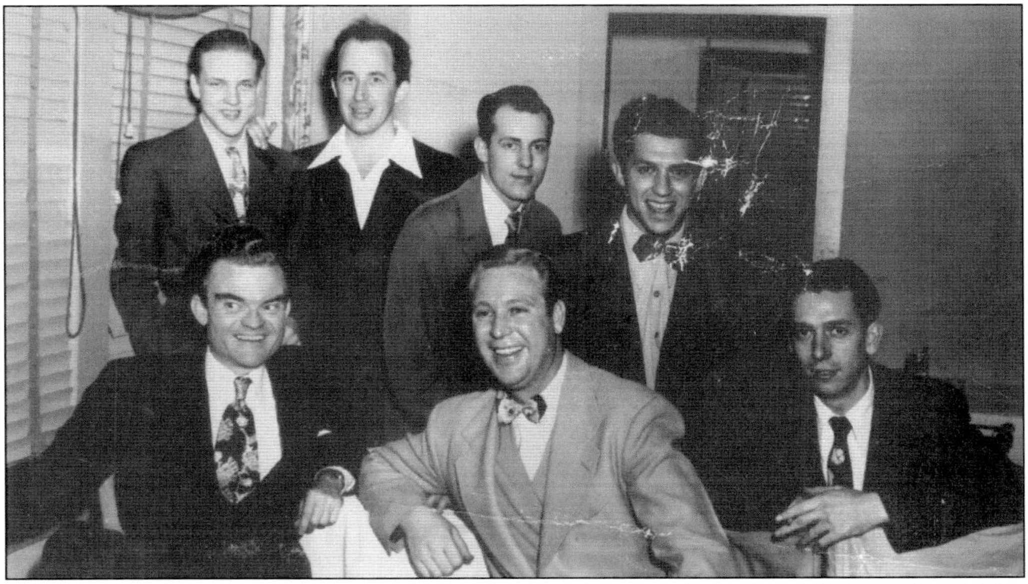

National funnyman and CBS Radio star Spike Jones poses with the WNOX house band. The recording artist is officially known as the creator of the on-air German dictator, a spoof spawned to give Adolph Hitler the old Bronx cheer. He performed in nearly every great radio comedy in the 1930s and 1940s. From left to right are (first row) Spike Jones and Eddie Hill, who would go on to perform on the Grand Ole Opry and be one of the first living inductees into the Country Music Disc Jockey Hall of Fame in 1975; (second row) Billy Bowman, John Gallaher, unidentified, Tony Cianciolo, and Troy Hatcher. (Courtesy of Harley Sweet.)

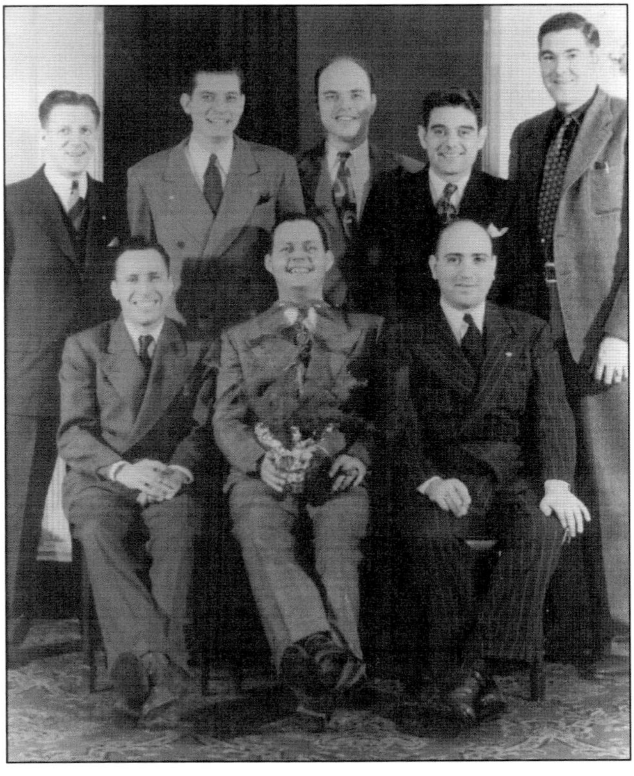

This is a group photograph taken at the 1945 WNOX Christmas party. From left to right are (first row) Bill Carlisle, Lowell Blanchard holding a poinsettia, and Tony Musco; (second row) Doc Johnson, Charles McMahon, two unidentified, and cowboy singer Homer Harris. (Courtesy of Harley Sweet.)

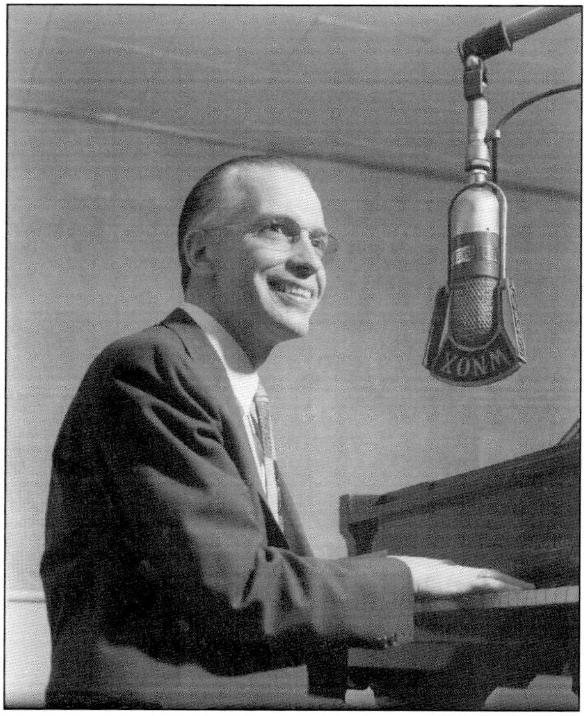

Jerry Collins toured as a pianist with national orchestras before coming to WNOX, where he worked as the music director of the station. He was part of the house band at the *Mid-Day Merry-Go-Round* and the *Tennessee Barn Dance*. (Courtesy of Harley Sweet.)

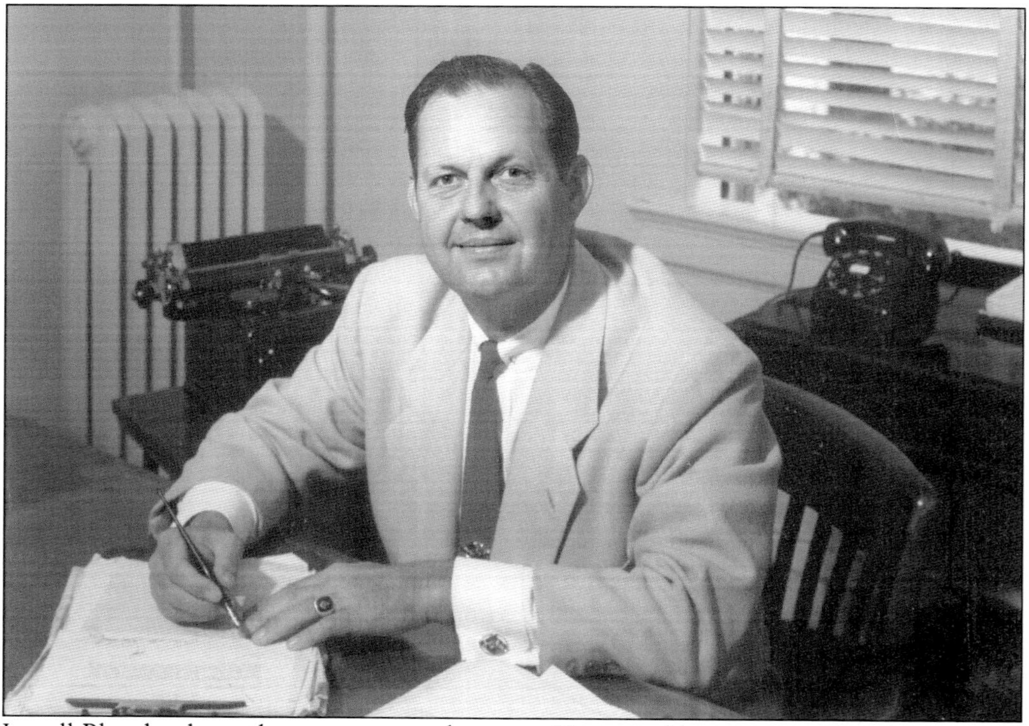

Lowell Blanchard was the consummate showman. He wrote his own material and improvised numerous skits on stage. He could deliver the punch line or play the straight man when called upon on stage with a relaxed, easy manner. Blanchard is pictured here, pen in hand, checking scripts for the day's show. (Courtesy of Harley Sweet.)

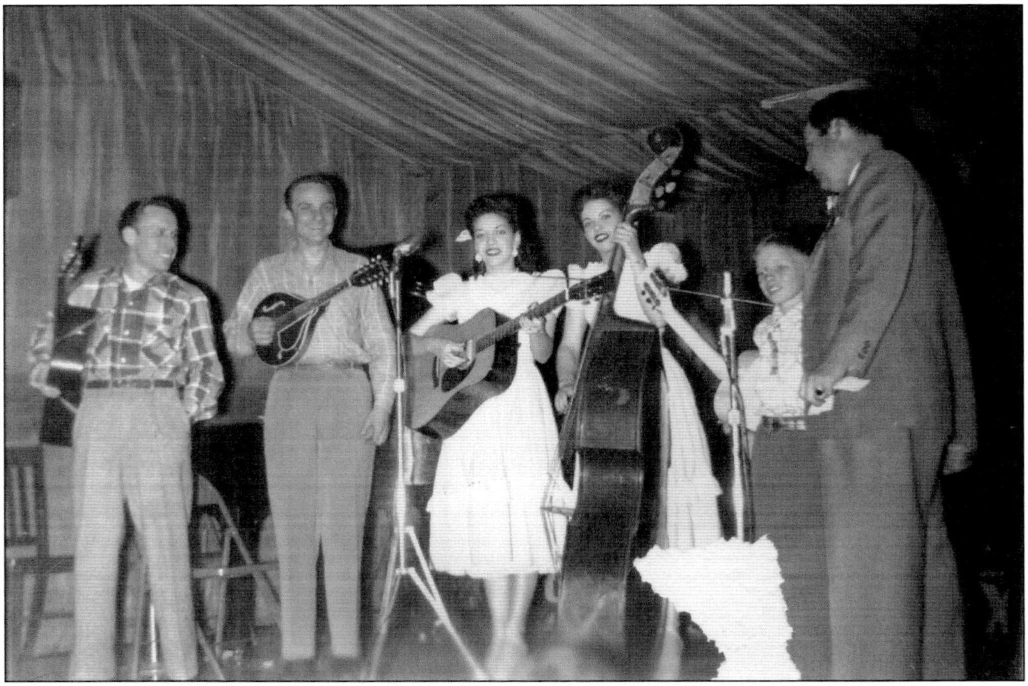

This is a cast of regulars who appeared on WNOX. From left to right are Bill Carlisle, James Carson, Martha Carson, Mattie O'Neal, an unidentified boy, and Lowell Blanchard. The cast on stage here were the program's most popular performers. Below is a photograph of Martha Carson (left) and Mattie O'Neal. O'Neal was a superb female vocalist and yodeler who would go on to rise to fame at the WSB Barn Dance in Atlanta. (Both courtesy of Roy Garrett.)

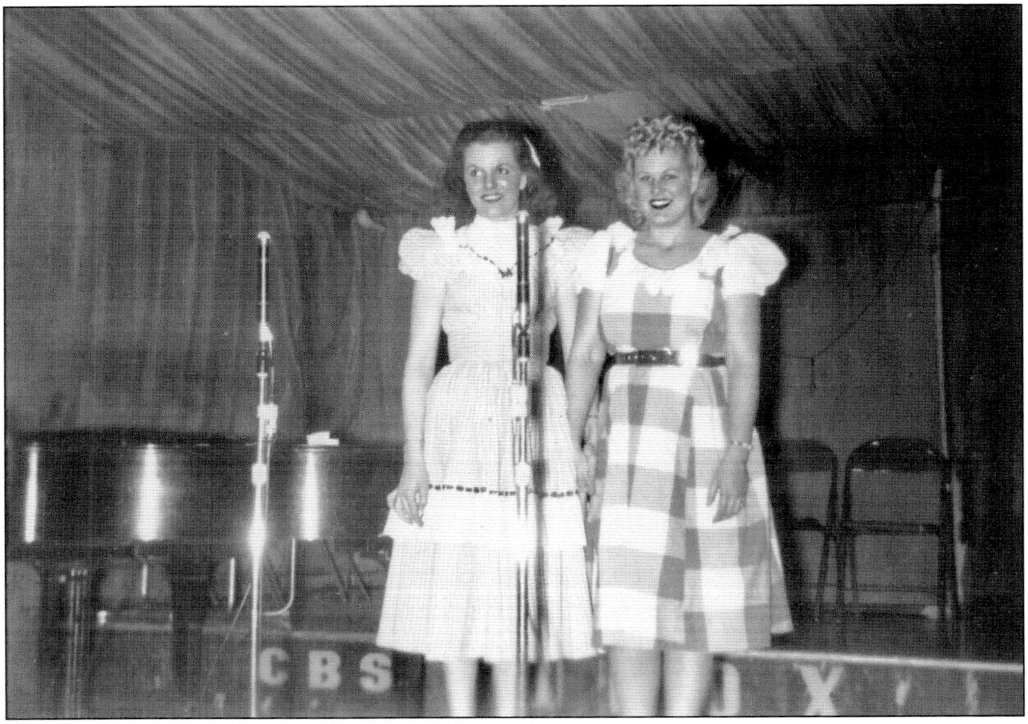

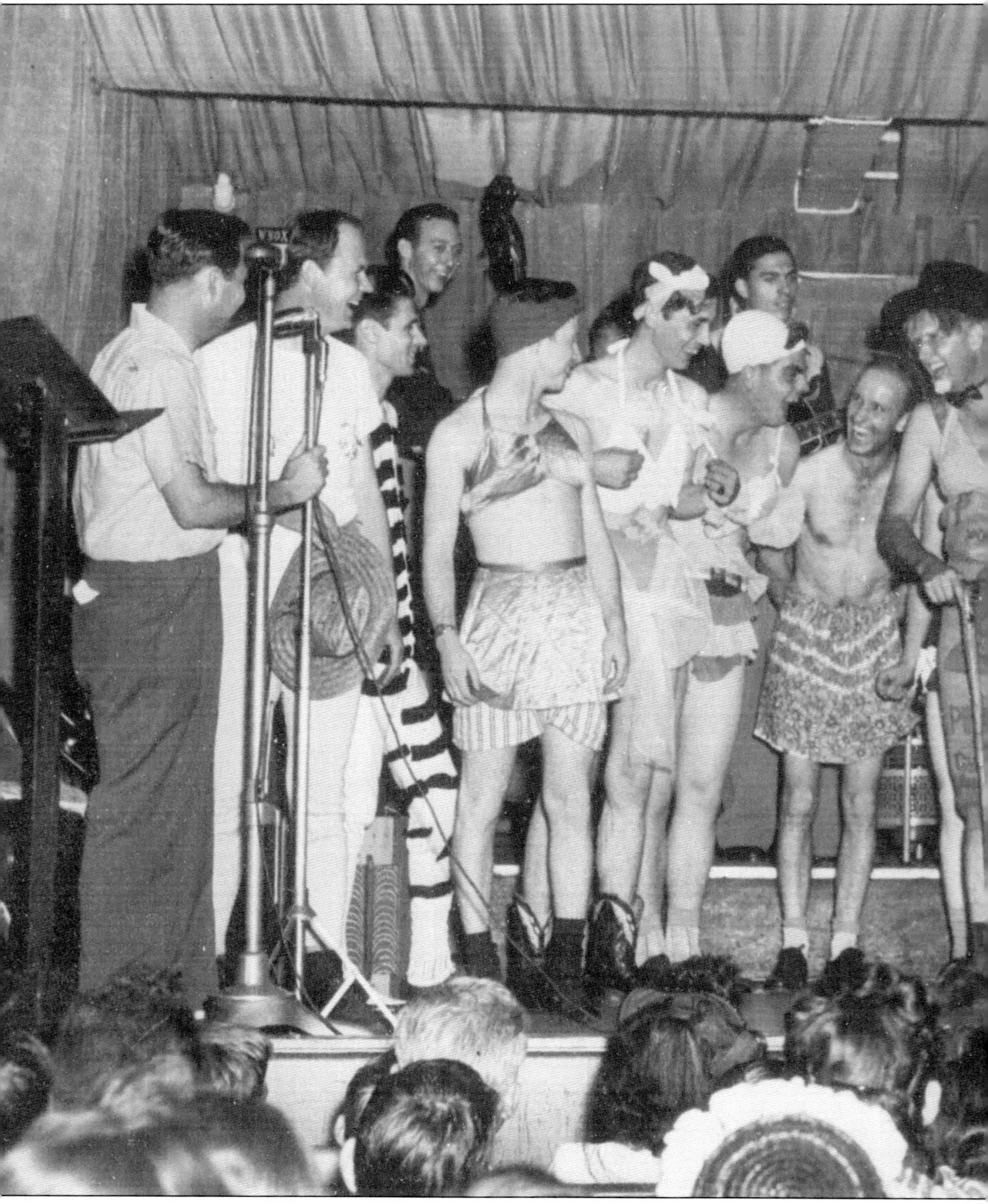

It was not only those listening at home whom Lowell Blanchard and crew had to keep entertained; the *Mid-Day Merry-Go-Round* and *Tennessee Barn Dance* had a live audience, and the programs were a major attraction in Knoxville, where people from around the region and across the nation would attend, eat at the restaurants, stay at the hotels, and shop at local businesses. Word of

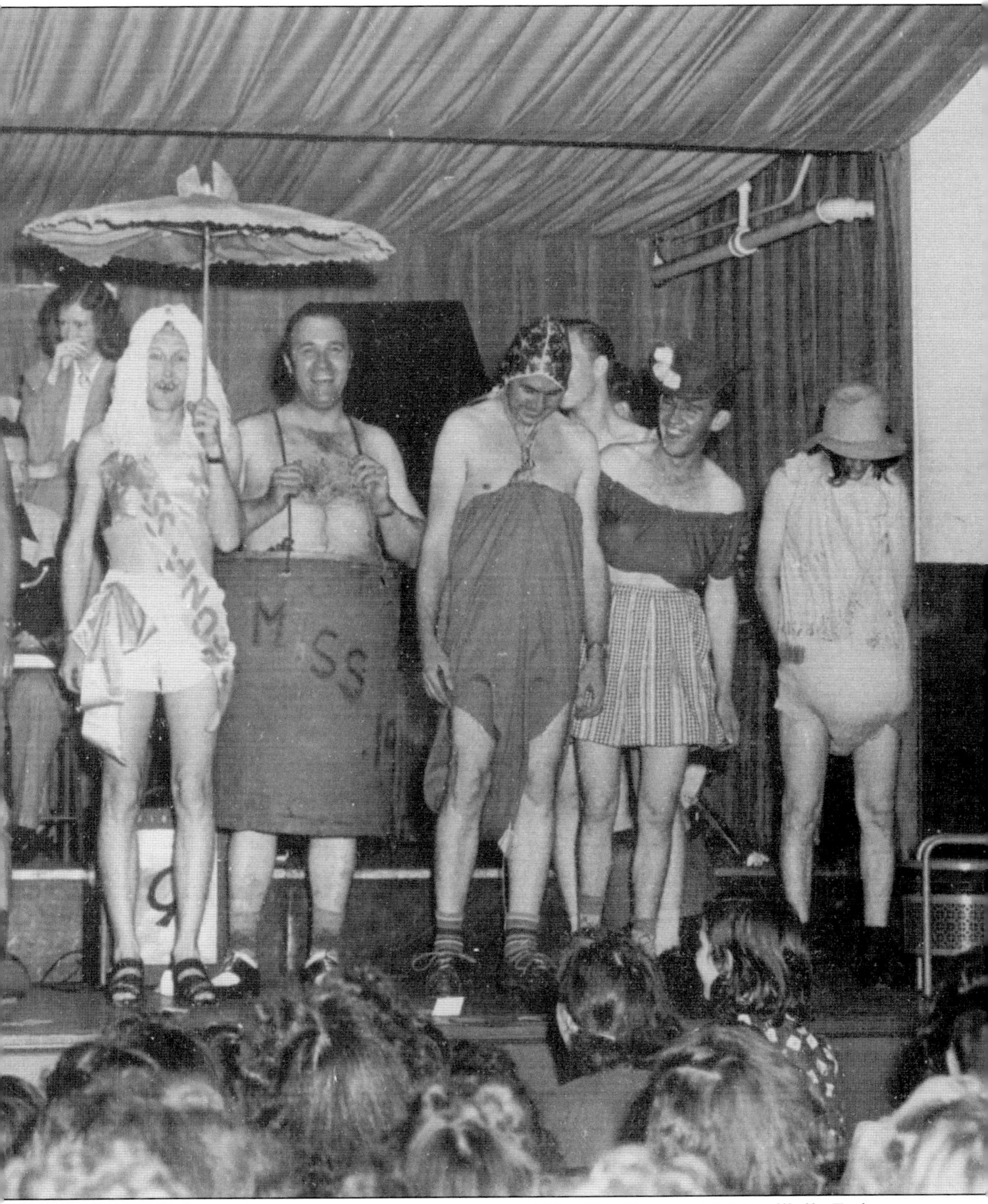

mouth was the best advertisement and promotion for WNOX. This is a 1945 Hillbilly Fashion Show that provided some much-needed comic relief in post–World War II Knoxville. At far left is announcer Lowell Blanchard. (Courtesy of Harley Sweet.)

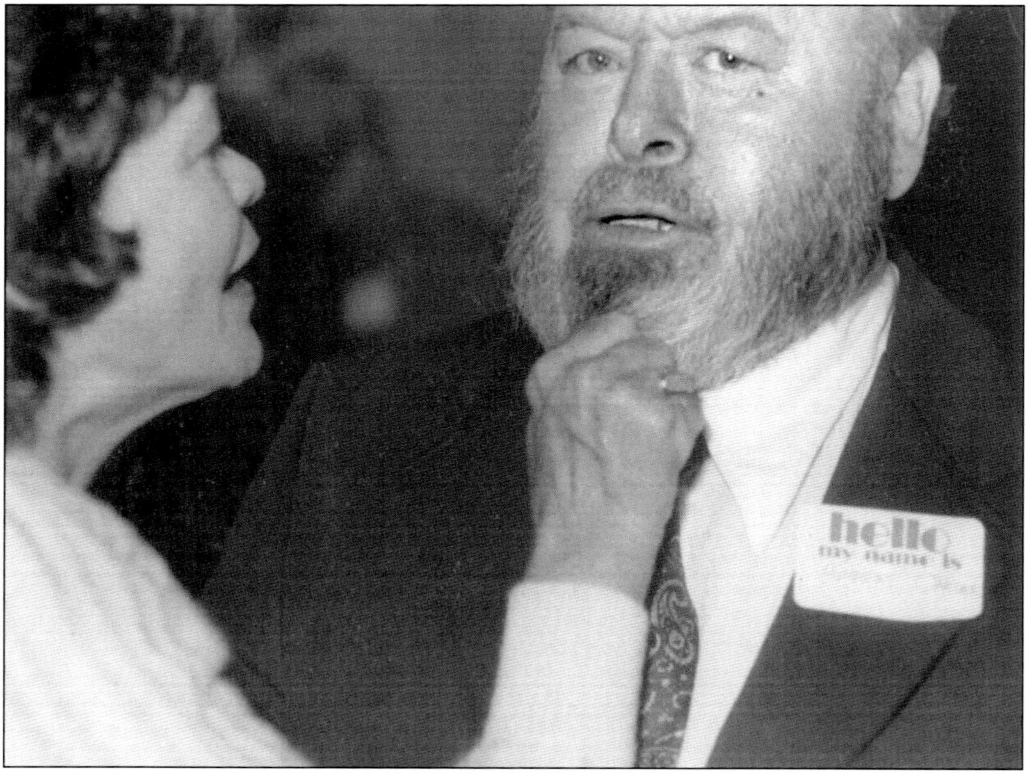

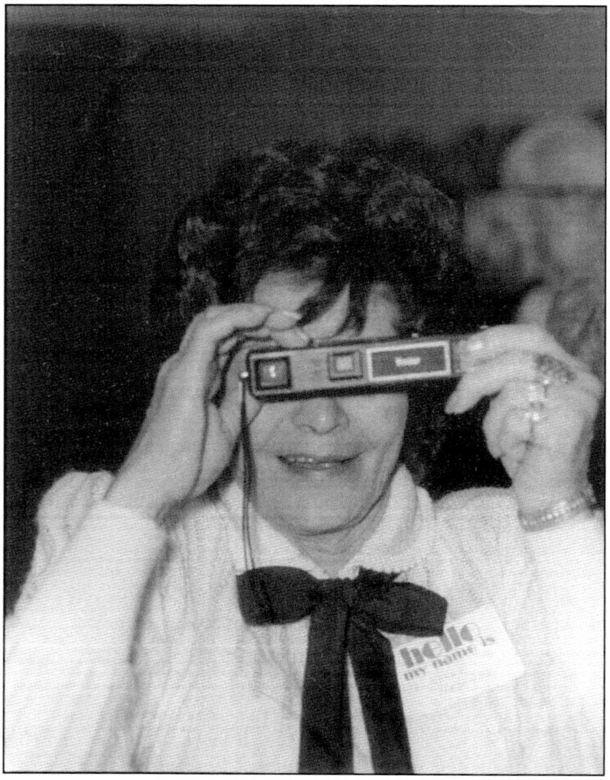

Knoxville native and actress Polly Bergen pulls on WNOX traffic director Hubert Carter's beard. Nellie Paulina Burgin grew up in the bluegrass community in Knoxville and made appearances on WNOX as a child musician. She started in Hollywood in lighthearted roles with Dean Martin and Jerry Lewis and worked her way to become a serious actress in movies like *Cape Fear* and *The Winds of War*. She is also a successful businesswoman and wrote three bestselling beauty books. (Courtesy of Harley Sweet.)

At left, she turns the camera on the photographer in this picture. Polly Bergen is still acting. Her latest most notable role is as the mother of Felicity Huffman's character, Lynette Scavo, in the American Broadcasting Company's television series *Desperate Housewives*. (Courtesy of Harley Sweet.)

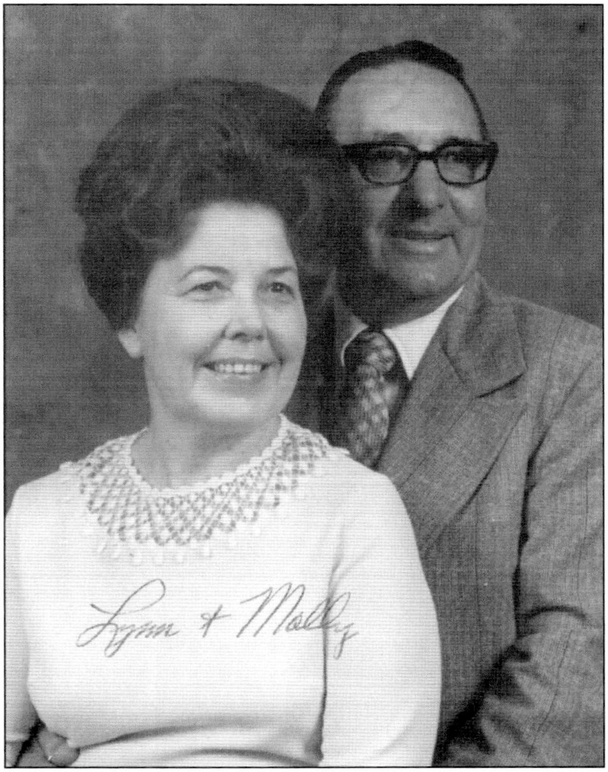

Molly O'Day, pictured here with her husband, Lynn Davis, is credited with pioneering the role of the solo female country artist. She was with WNOX for five years and was one of the most popular entertainers on the air. She was signed to a Columbia Recording contract, and Fred Rose of Acuff Rose Music had her cut eight songs written by Hank Williams that increased her popularity tremendously. In 1950, she and her husband quit the road to sing in churches, and Davis joined the ministry. She continued recording for Columbia and other smaller labels. (Courtesy of Harley Sweet.)

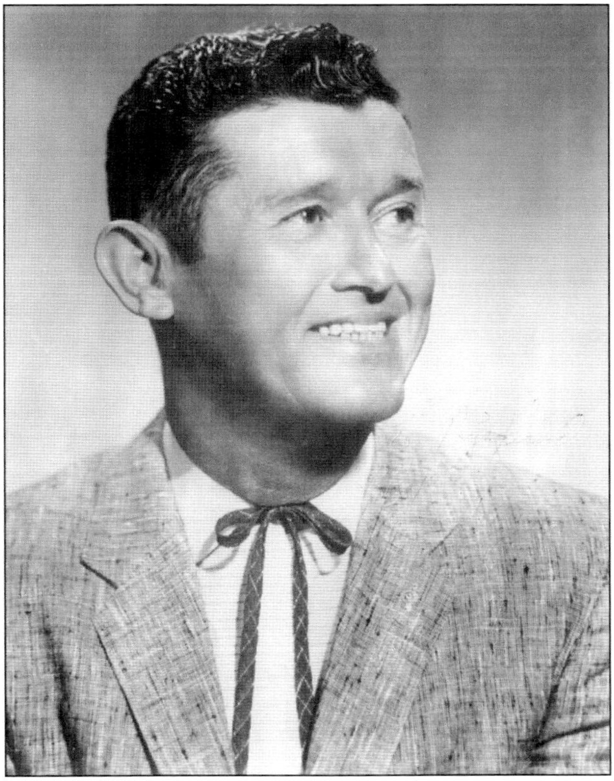

Maynardville native Roy Acuff is regarded as the "King of Country Music" and the first millionaire in the industry. He was a graduate of Central High School and sought a career in professional baseball before turning to music. He started on the old Medicine Show circuit and learned to sing loud enough that, when he transitioned to radio, he could dominate the band with his voice. He played at WNOX prior to moving to the Grand Ole Opry, where he became one of its most popular performers. In 1942, he and friend Fred Rose founded Acuff Rose Music. (Courtesy of Roy Garrett.)

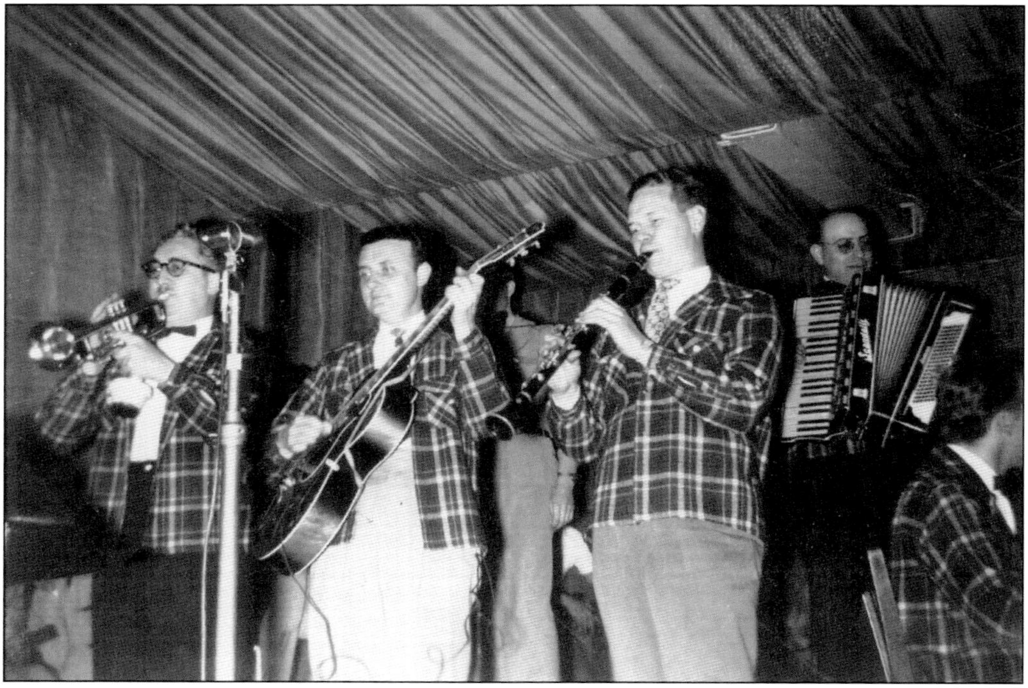

It wasn't all banjos, fiddles, and guitars. Here the WNOX house band does a warm-up number between the sets. Above from left to right are Dave Dunham on trumpet, Charlie Hagaman on guitar, Buck Houtchins on clarinet, and Tony Musco on accordion. Comedy was always part of the act in every band's stage patter. Below is Hot Shot Elmer (left, also known as Bill Carlisle) and pianist Jerry Collins, who also served as WNOX's music director. (Both courtesy of Roy Garrett.)

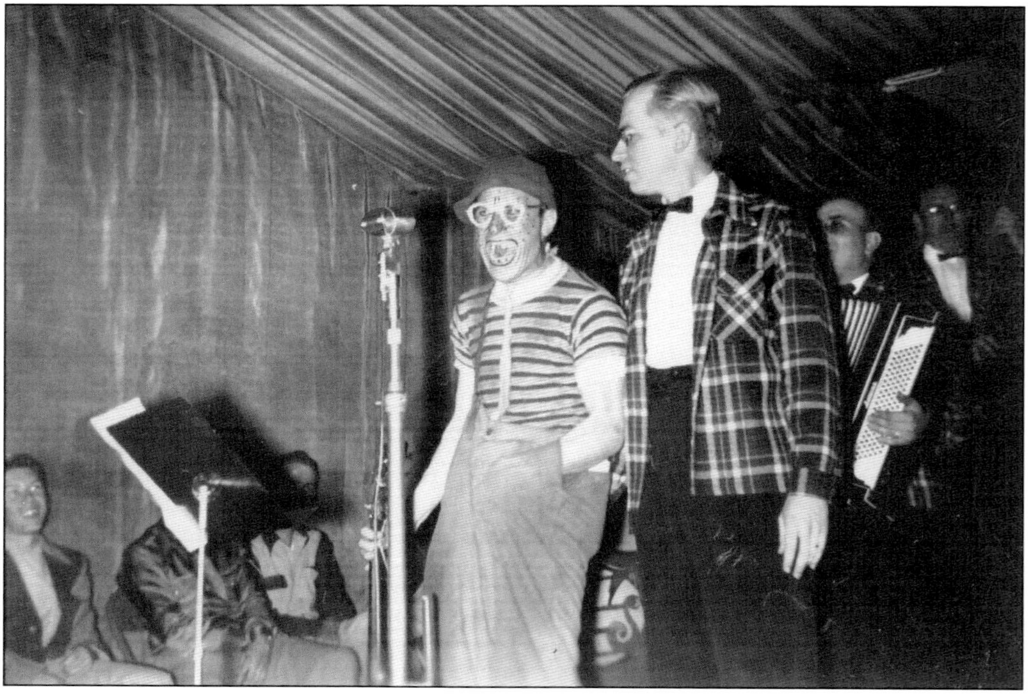

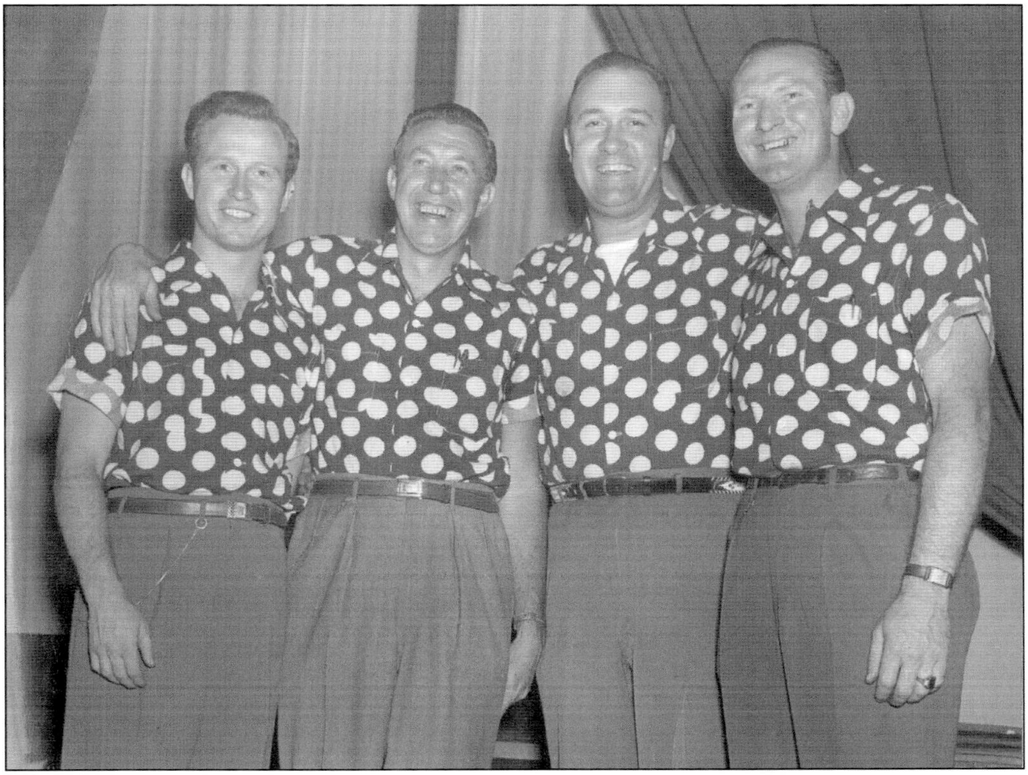

This is a promotional photograph of Carl Story and the Rambling Mountaineers. From left to right are Red Rector, Claude Boone, Carl Story, and Cotton Gaylon. They were formidable musicians and entertained crowds at the *Mid-Day Merry-Go-Round* and *Tennessee Barn Dance*. (Courtesy of Harley Sweet.)

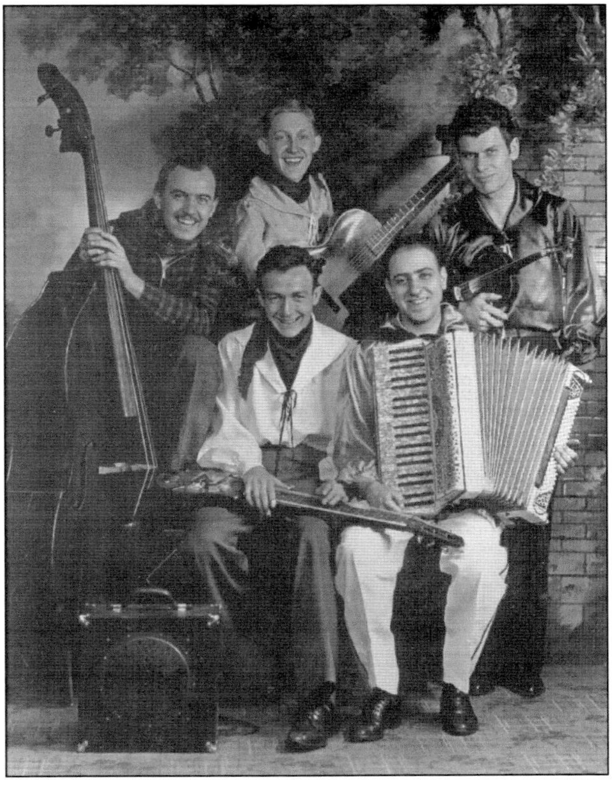

Mid-Day Merry-Go-Round musicians pose for a photograph in this image. From left to right are (first row) Tony Musco and Guy Campbell; (second row) Archie Campbell, Homer Haynes, and Tommy Covington. (Courtesy of Harley Sweet.)

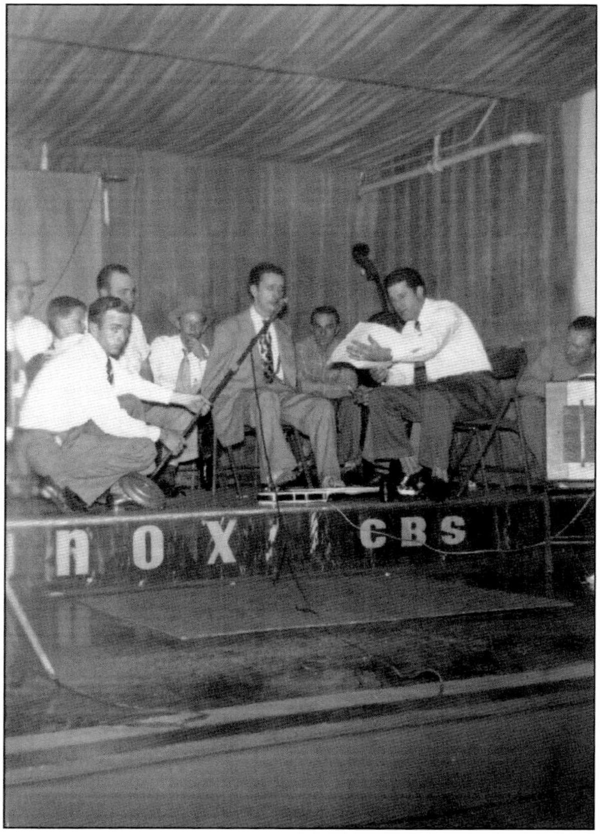

WNOX featured a talented individual named Ray Meyers. He was a gifted vocalist without arms who played the guitar with his feet. Ray is pictured here with one man holding the music while another holds the microphone. (Courtesy of Roy Garrett.)

Ira and Charlie Loudermilk changed their names to the Louvin Brothers when they entered music. Here they are pictured on stage at the *Mid-Day Merry-Go-Round.* Ira Louvin is on mandolin and Charlie on guitar. They made close harmony vocals popular on stage. They went on to a popular singing career in Nashville and were inducted into the Country Music Hall of Fame in 2001. (Courtesy of Roy Garrett.)

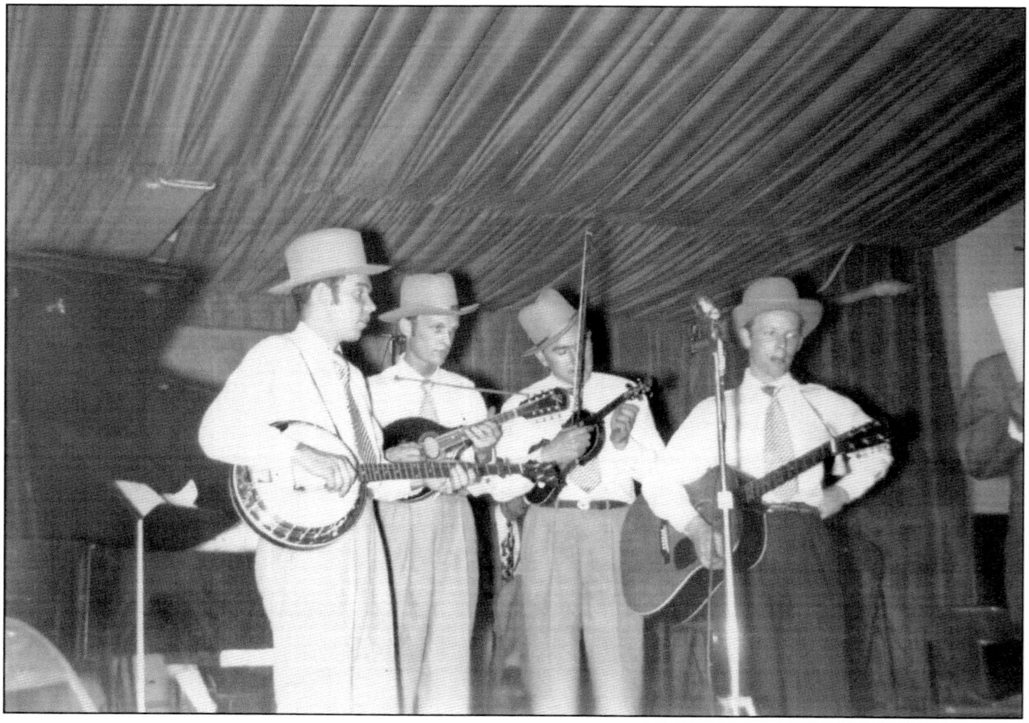

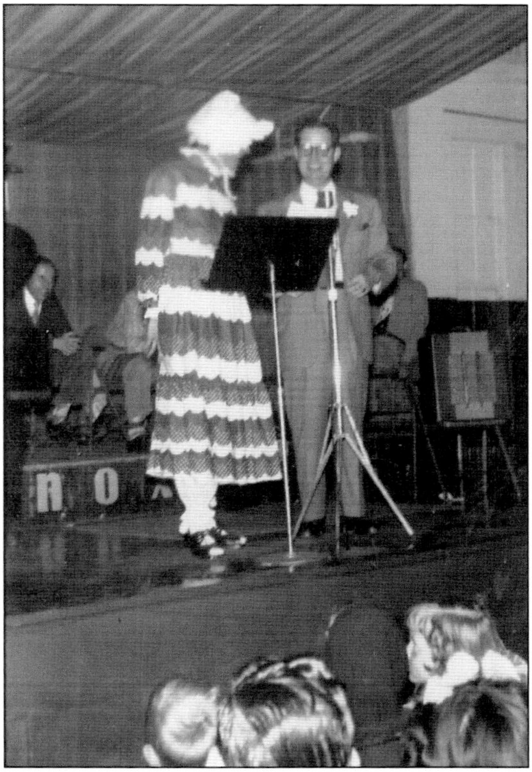

Ira Louvin was a cutup and is shown here in drag portraying Sal Skinner in a comedy sketch for the program with Lowell Blanchard. His and his brother's foray into comedy would result in some popular recordings in later years in Nashville. Below is an image of "Old Joe Clark," who was a regular comedian on the *Mid-Day Merry-Go-Round* and the *Tennessee Barn Dance*. (Both courtesy of Roy Garrett.)

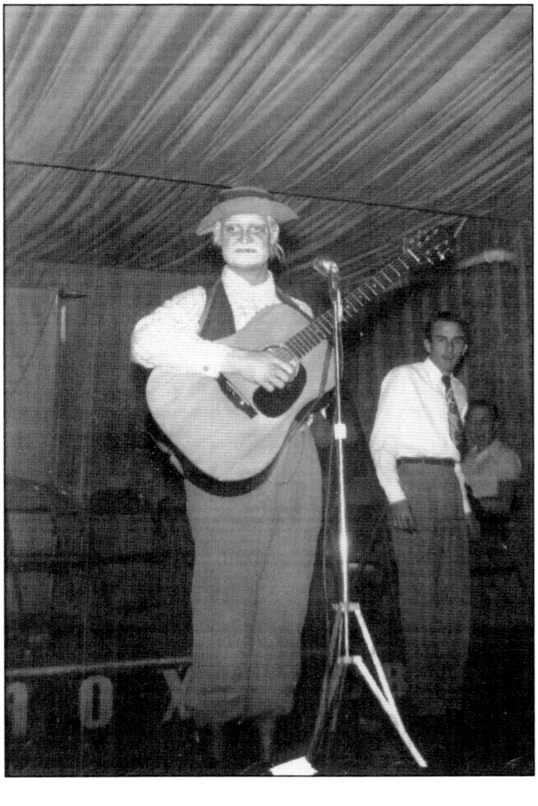

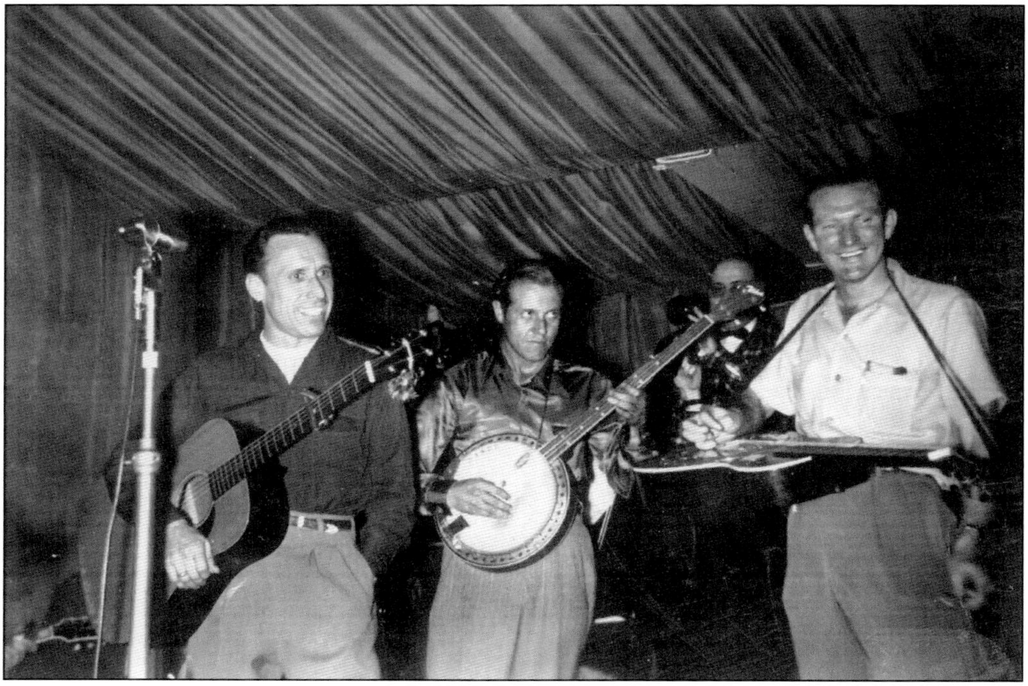

This is a photograph of Bill Carlisle (far left) with an unidentified banjo player and Cotton Gaylon on Dobro. When he wasn't playing "Hot Shot Elmer," Bill Carlisle was a serious musician and gifted guitar player. He always maintained close ties with WNOX throughout his career. Carlisle was an excellent yodeler, an entertainer who scored Top 10 hits, and a good songwriter, penning the popular 1970s *Hee Haw* gospel quarter signature song "Gone Home." He was inducted into the Country Music Hall of Fame in 2002 and performed until his death on March 17, 2003. (Courtesy of Roy Garrett.)

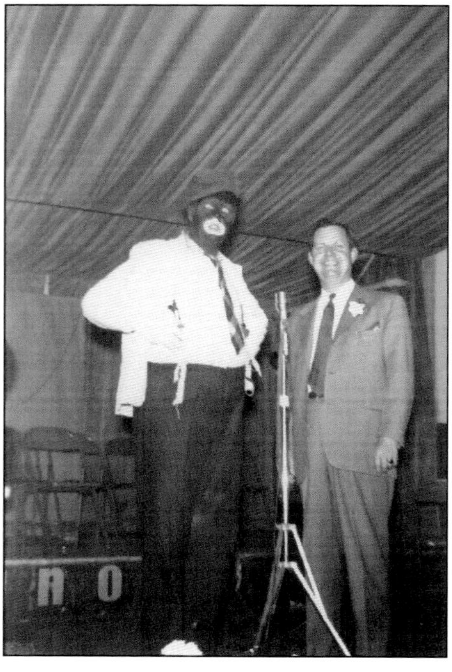

WNOX announcer Lowell Blanchard and "Kentucky Slim" (Charles Elza) do an on-air skit called "Ghosts Strike at Midnight." Blanchard was a student of vaudeville and collected many books on the subject, substituting rural humor for the city humor often used in comedy acts. (Courtesy of Roy Garrett.)

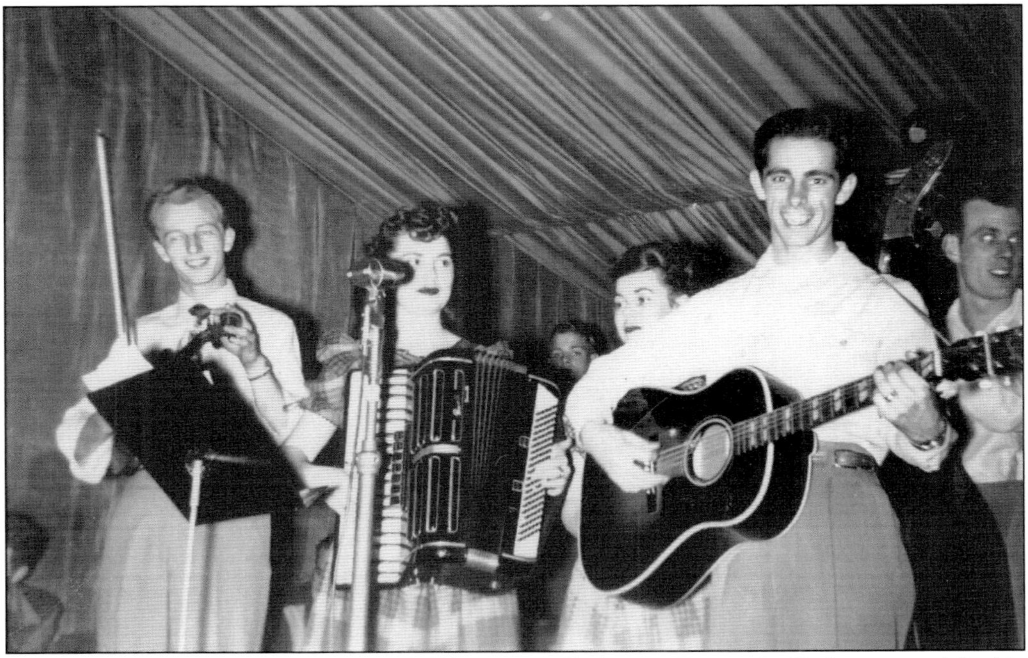

From left to right are Buck Ryan on fiddle, Bea Puffenbarger, Bunny Puffenbarger, and Buddy Puffenbarger. Not every band featured went on to fame in Nashville. Many preferred playing at community events, fairs, and churches away from the big crowds. Lowell Blanchard, however, was a master at getting them all to WNOX's stages to perform. (Courtesy of Roy Garrett.)

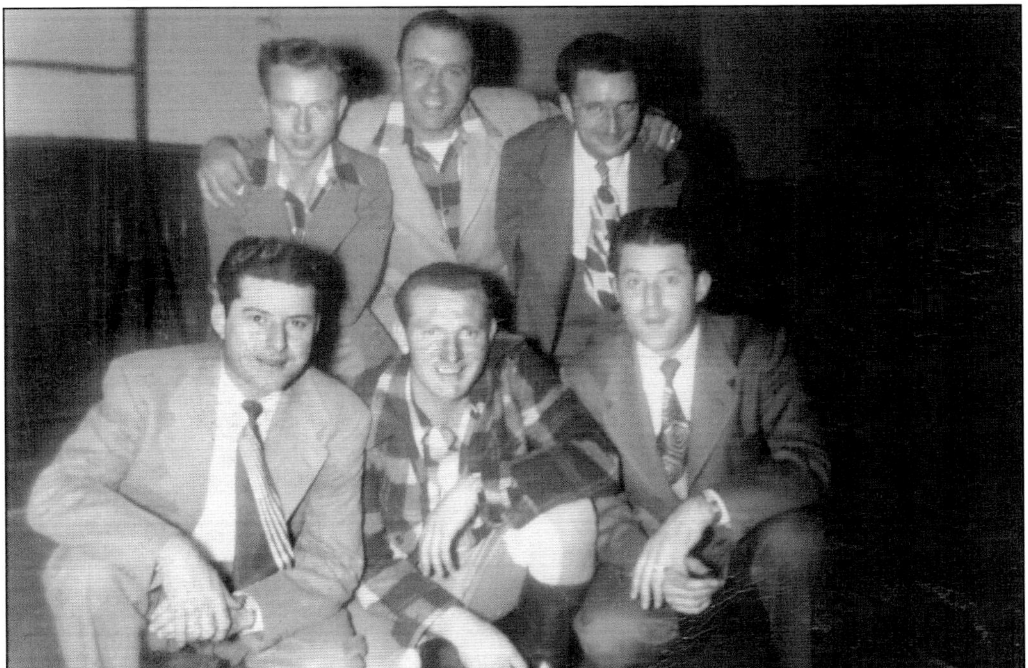

This photograph of Carl Story and the gang was taken by a fan at the *Mid-Day Merry-Go-Round*. Pictured here from left to right are (first row) Hack Johnson, Cotton Gaylon, and Clyde Johnson; (second row) Red Rector, Carl Story, and Ray Meyers. (Courtesy of Roy Garrett.)

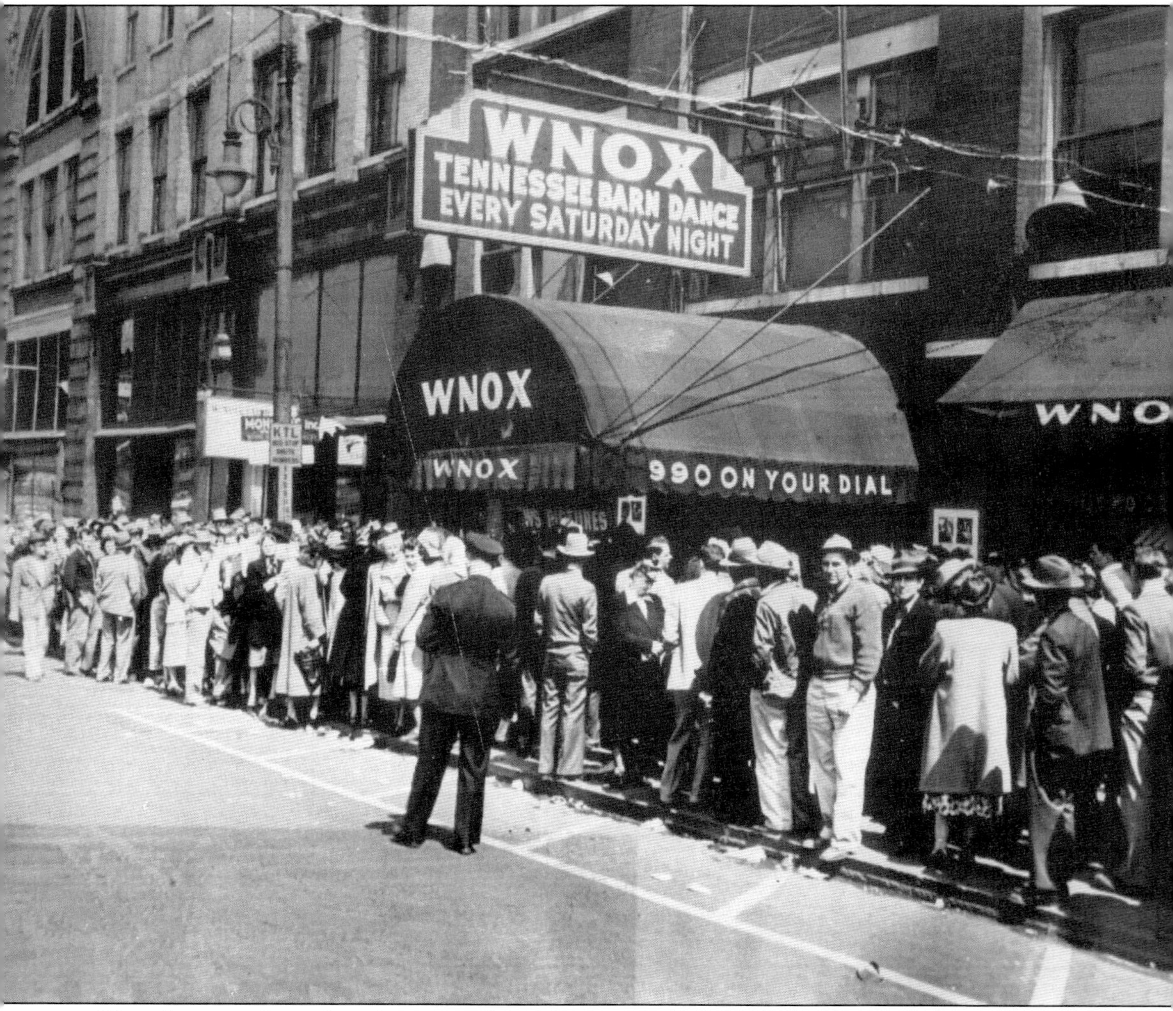

This photograph is taken facing north on Gay Street showing the crowds waiting for the doors to open to the *Mid-Day Merry-Go-Round*. The radio program and the *Tennessee Barn Dance* helped establish an entire musical community in the city. (Courtesy of Harley Sweet.)

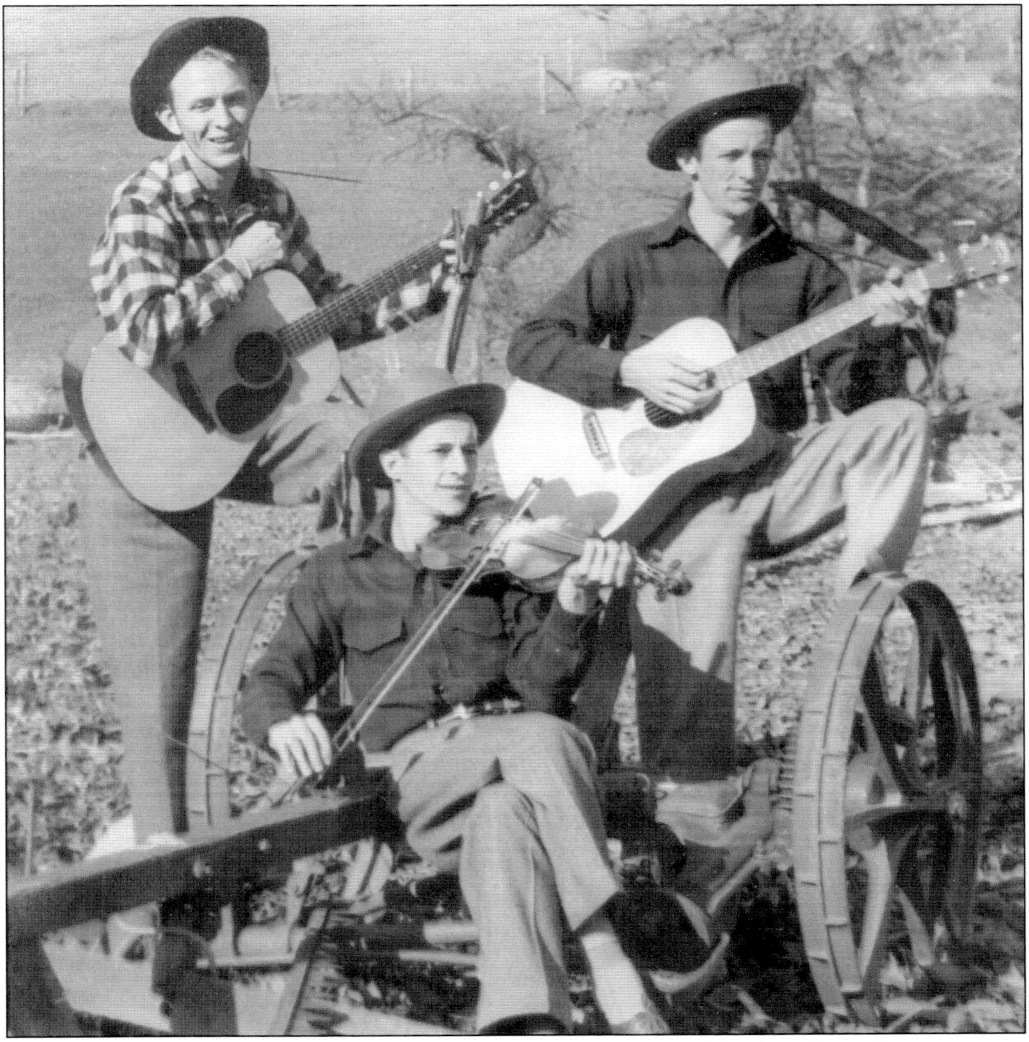

Pictured here is Harley "Sunshine Slim" Sweet (left) and the Burchfield Brothers. Sweet was one of the most prolific musicians. He joined the Burchfields and left soon after to join Molly O'Day. It was not uncommon for musicians at WNOX to vary who they played with from time to time. (Courtesy of Harley Sweet.)

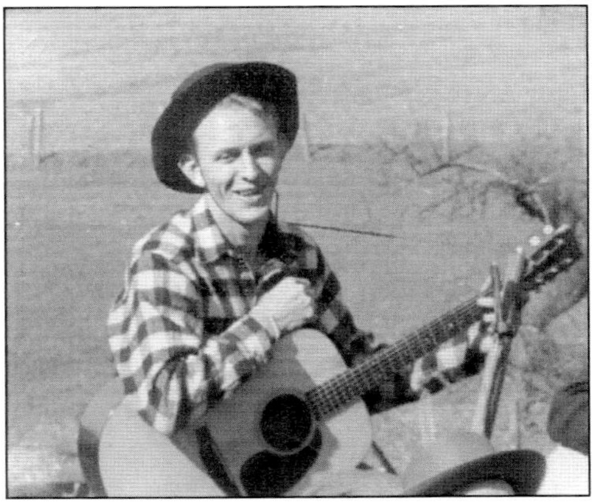

Sweet, like many of WNOX's musicians and staff, took an unplanned break from music during World War II. He volunteered for the U.S. Navy and worked as a Seabee. During his time in the South Pacific, he saw action in numerous campaigns building beach heads, including the Solomon Islands Campaign. He is the only surviving musician left from the *Mid-Day Merry-Go-Round* today who still plays. (Courtesy of Harley Sweet.)

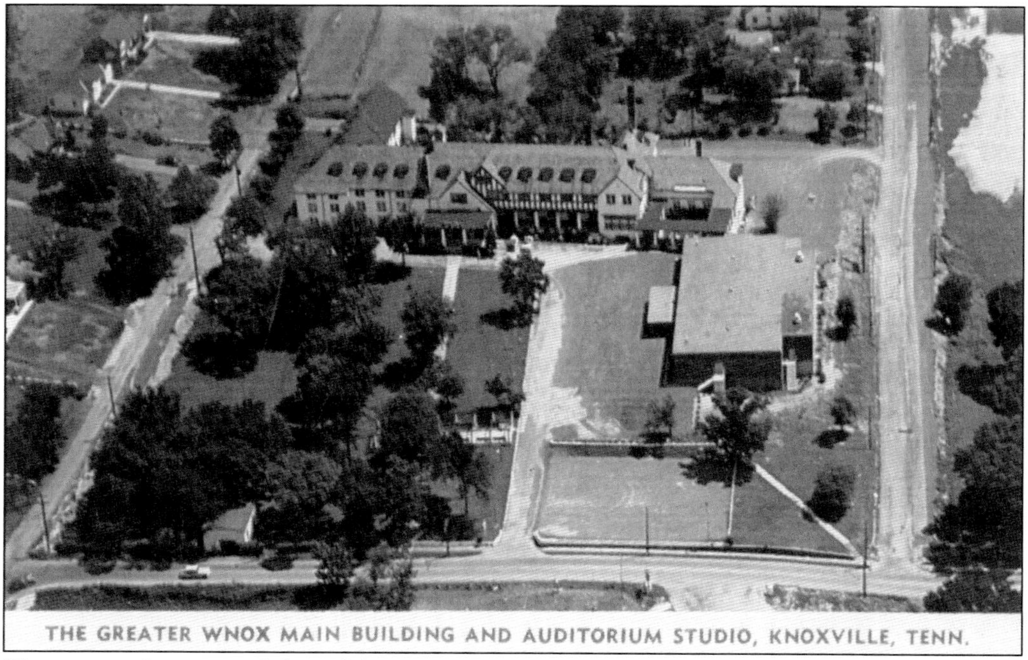

This postcard is an aerial shot of the WNOX Auditorium Studio and Whittle Springs Hotel that was dedicated May 12, 1955, in North Knoxville. The former resort property was purchased to accommodate both a radio and a television station. When the television license was not issued to Scripps Howard, the entire facility was converted over to radio and it became the only radio station studio of its kind in the world. The auditorium seated 1,250 people on a 6-acre piece of property.

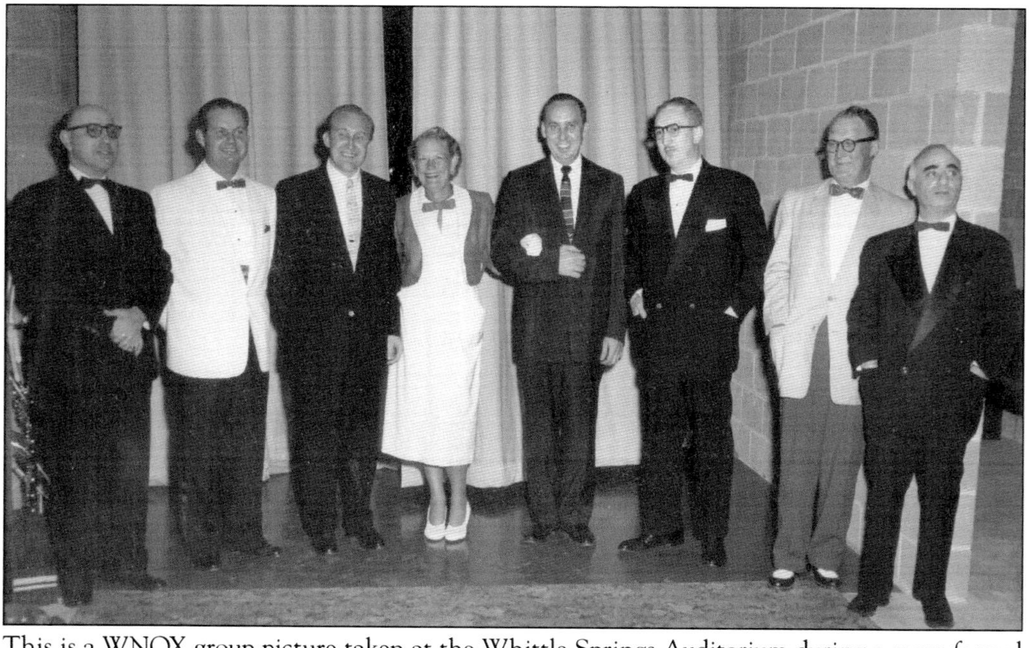

This is a WNOX group picture taken at the Whittle Springs Auditorium during a more formal moment. From left to right are Tony Musco, Lowell Blanchard, Homer Haynes, Mrs. Blanchard (Sally), Jethro Burns, Dave Durham, O. L. Smith, and Harry Nides. (Courtesy of Harley Sweet.)

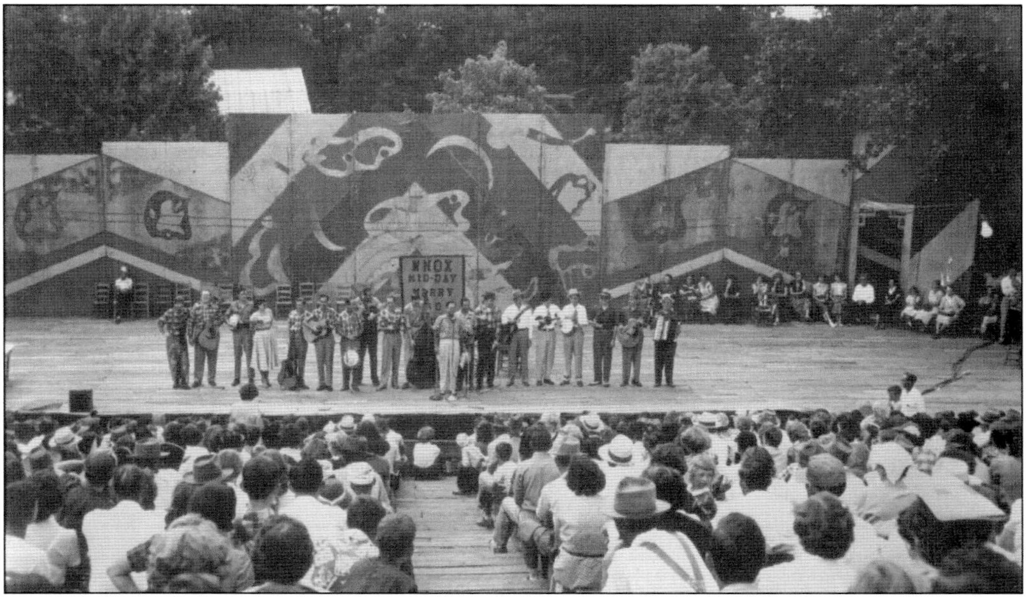

The WNOX *Mid-Day Merry-Go-Round* became an annual program at the Tennessee Valley Agricultural and Industrial Fair. The station set the standard for bringing big-name acts to Knoxville every fall for the event. Here Lowell Blanchard is at the microphone announcing the next act.

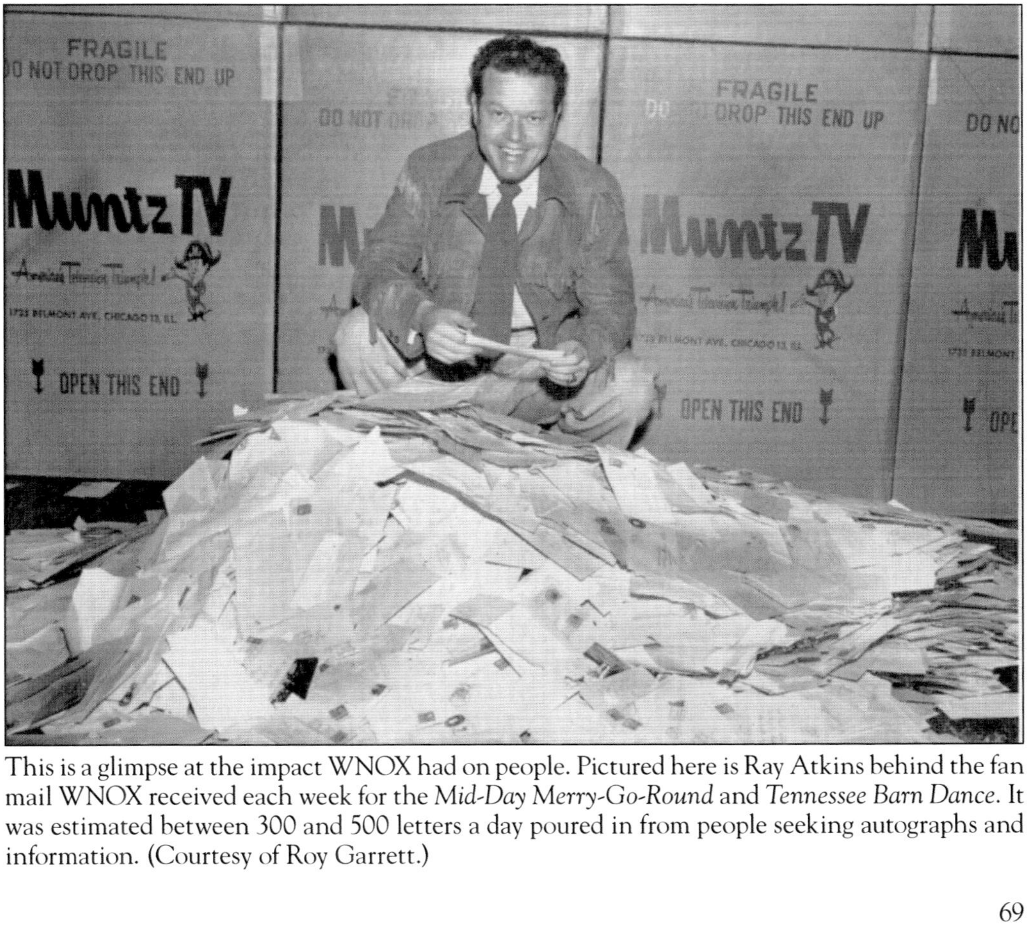

This is a glimpse at the impact WNOX had on people. Pictured here is Ray Atkins behind the fan mail WNOX received each week for the *Mid-Day Merry-Go-Round* and *Tennessee Barn Dance*. It was estimated between 300 and 500 letters a day poured in from people seeking autographs and information. (Courtesy of Roy Garrett.)

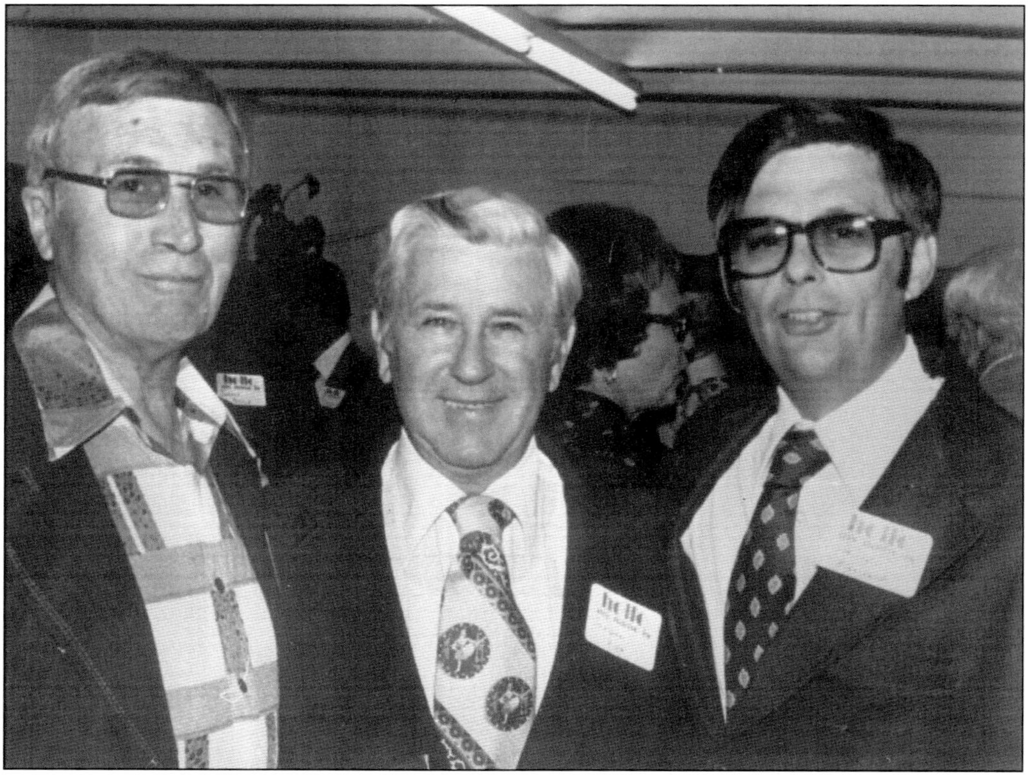

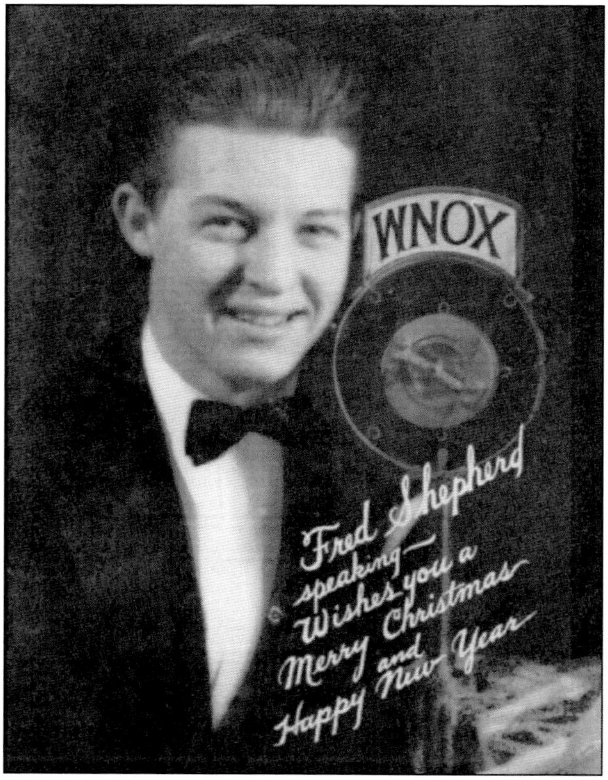

WNOX was a popular place for country music entertainers and songwriters to revisit once they made it in the industry. Pictured above from left to right are songwriter Mel Foree; Pee Wee King, who first recorded the song "The Tennessee Waltz" that became one of the official songs of the state; and Smiley Blanchard, the son of WNOX announcer Lowell Blanchard. (Courtesy of Harley Sweet.)

This is Fred Shepherd's *c.* 1940 WNOX Christmas card. It is the only known surviving card of the ones station personnel sent to friends and family as a holiday promotion. (Courtesy of the Fred Shepherd family.)

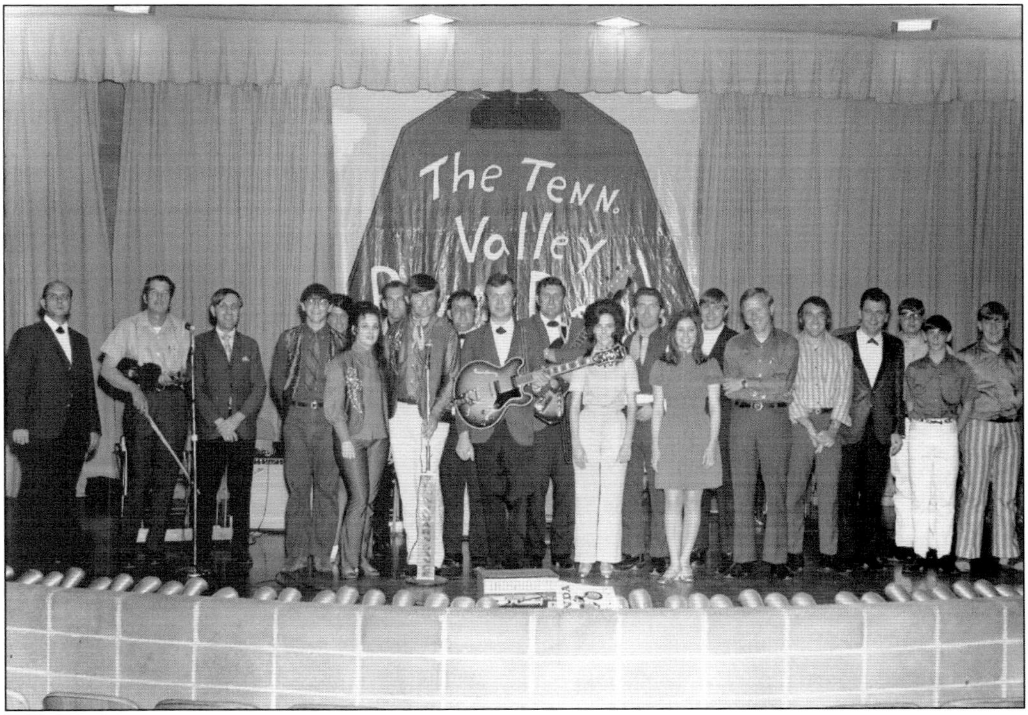

After WNOX made the switch to a Top 40 format, the *Tennessee Barn Dance* added the word "Valley" to avoid trademark complications and leased the auditorium to continue holding them. This is a photograph taken in 1971 of the cast of the *Tennessee Valley Barn Dance*. Below, a singer entertains a sizeable audience of schoolchildren. The WNOX auditorium was still a popular place for field trips and programs. (Both courtesy of John Hitch.)

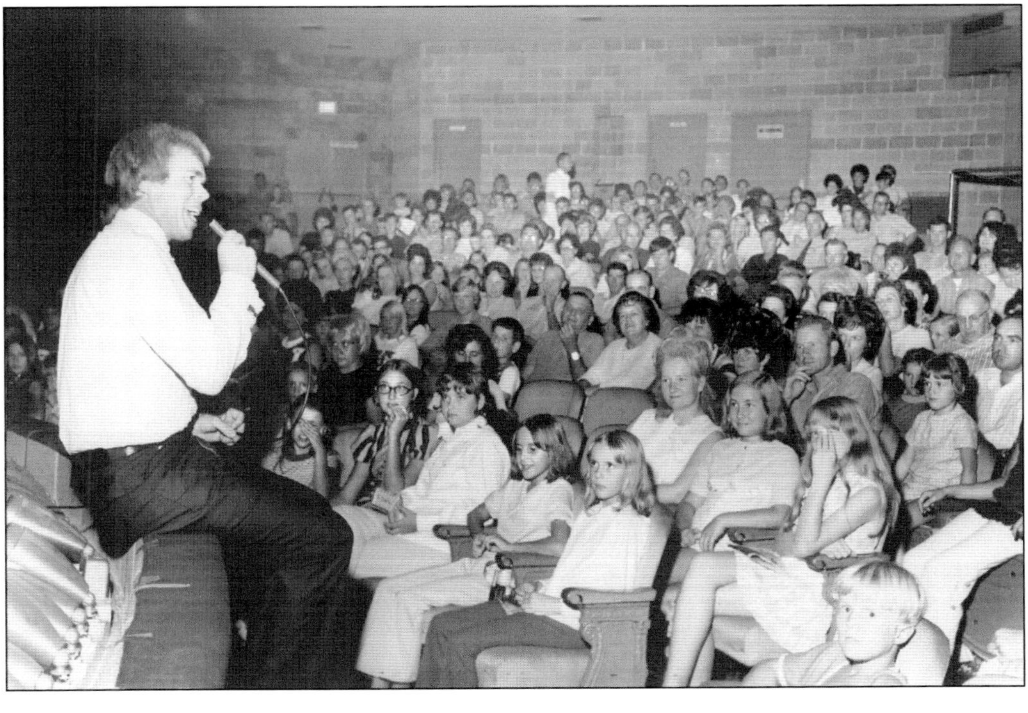

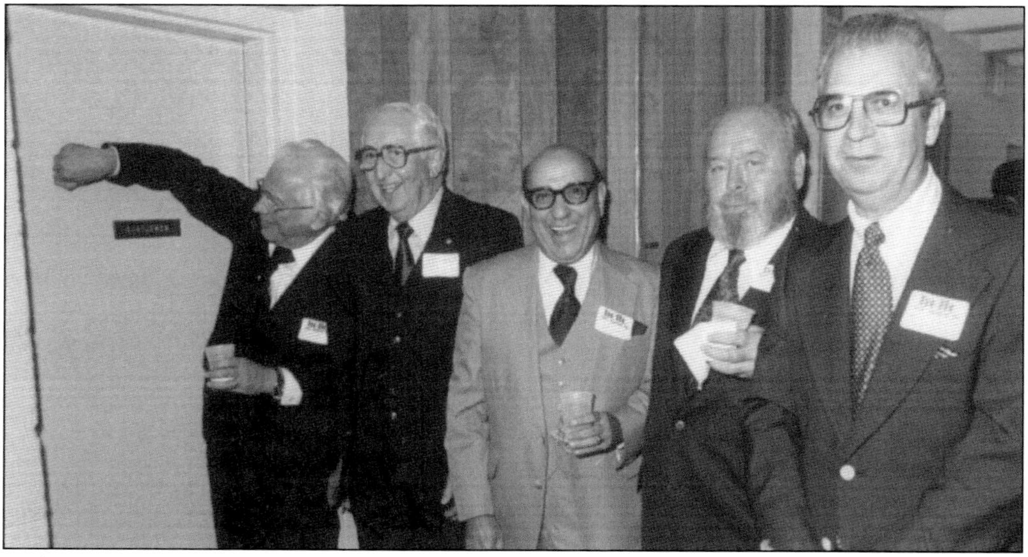

Mid-Day Merry-Go-Round reunions were always a popular event attended by former musicians, who remember being a very close-knit group. This comical photograph was taken at the 1987 reunion. From left to right are Jerry Collins, Dave Durham, Tony Musco, Hubert Carter, and Troy Hatcher. Below is "Sunshine Slim Sweet," who participates in the *Merry-Go-Round* reunions. At the writing of this book, he was 90 years old and one of only two remaining members of the WNOX programs. (Both courtesy of Harley Sweet.)

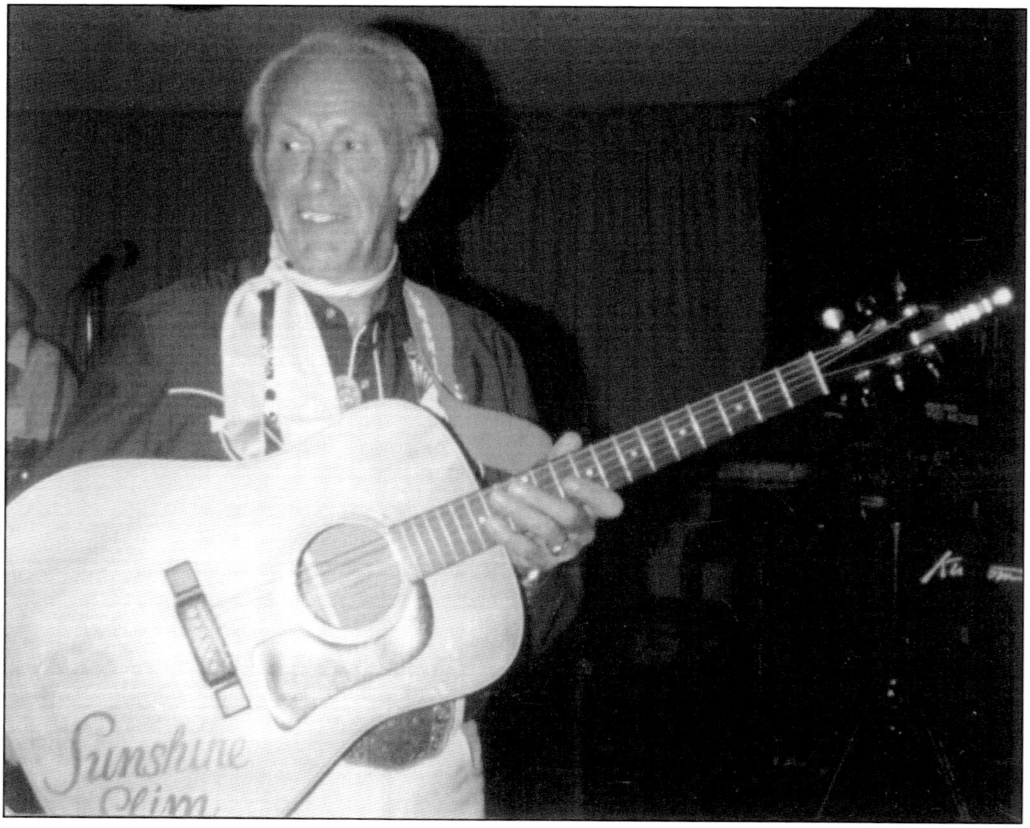

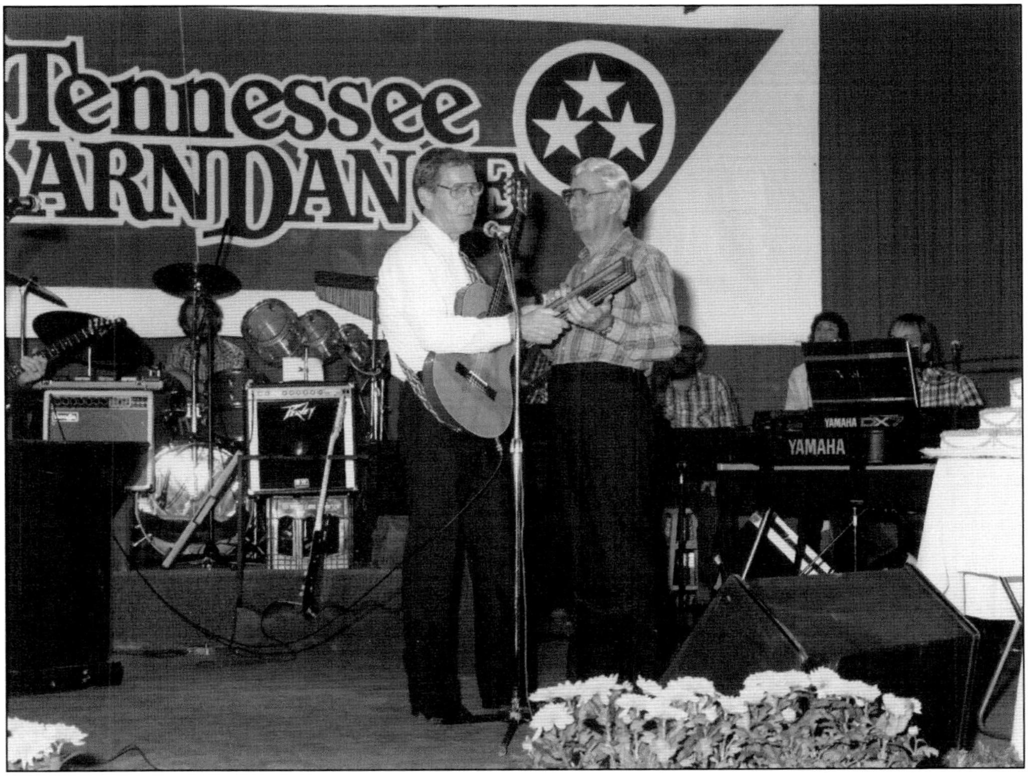

In 1985, WNOX had returned to a country music format and relaunched the *Barn Dance* with a special dedication ceremony. Here are country music entertainment great Chet Atkins (left) and Archie Campbell (right), who was master of ceremonies at the event. Below is the program from the event, which boasted a number of Nashville country music stars. (Both courtesy of Edith Kirkland.)

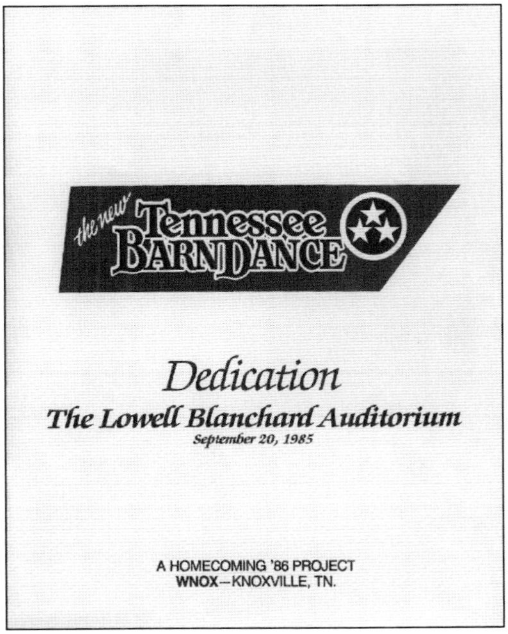

the new Tennessee BARNDANCE

Dedication

The Lowell Blanchard Auditorium

September 20, 1985

A HOMECOMING '86 PROJECT
WNOX—KNOXVILLE, TN.

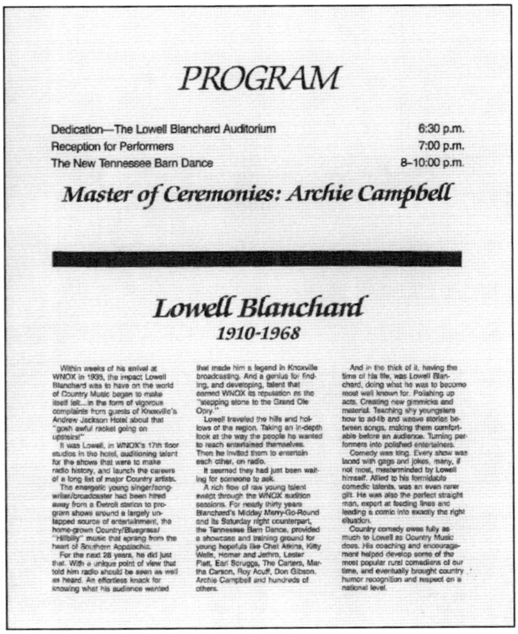

PROGRAM

Dedication—The Lowell Blanchard Auditorium	6:30 p.m.
Reception for Performers	7:00 p.m.
The New Tennessee Barn Dance	8–10:00 p.m.

Master of Ceremonies: Archie Campbell

Lowell Blanchard

1910-1968

Within weeks of his arrival at WNOX in 1935, the impact Lowell Blanchard was to have on the world of Country Music began to make itself felt...in the form of vigorous complaints from guests of Knoxville's Andrew Jackson Hotel about that "gosh awful racket going on upstairs!"

It was Lowell, in WNOX's 17th floor studios in the hotel, auditioning talent for the shows that were to make radio history, and launch the careers of a long list of major Country artists.

The energetic young singer/songwriter/broadcaster had been hired away from a Detroit station to program shows around a largely untapped source of entertainment, the home-grown Country/Bluegrass/"Hillbilly" music that sprang from the heart of Southern Appalachia.

For the next 28 years, he did just that. With a unique point of view that told him radio should be seen as well as heard. An effortless knack for knowing what his audience wanted that made him a legend in Knoxville broadcasting. And a genius for finding, and developing, talent that earned WNOX its reputation as the "stepping stone to the Grand Ole Opry."

Lowell traveled the hills and hollows of the region. Taking an in-depth look at the way the people he wanted to reach entertained themselves. Then he invited them to entertain each other, on radio.

It seemed they had just been waiting for someone to ask.

A rich flow of raw young talent swept through the WNOX audition sessions. For nearly thirty years Blanchard's Midday Merry-Go-Round and its Saturday night counterpart, the Tennessee Barn Dance, provided a showcase and training ground for young hopefuls like Chet Atkins, Kitty Wells, Homer and Jethro, Lester Flatt, Earl Scruggs, The Carters, Martha Carson, Roy Acuff, Don Gibson, Archie Campbell and hundreds of others.

And in the thick of it, having the time of his life, was Lowell Blanchard, doing what he was to become most well known for. Polishing up acts. Creating new gimmicks and material. Teaching shy youngsters how to ad-lib and weave stories between songs, making them comfortable before an audience. Turning performers into polished entertainers.

Comedy was king. Every show was laced with gags and jokes, many, if not most, masterminded by Lowell himself. Allied to his formidable comedic talents, was an even rarer gift. He was also the perfect straight man, expert at feeding lines and leading a comic into exactly the right situation.

Country comedy owes fully as much to Lowell as Country Music does. His coaching and encouragement helped develop some of the most popular rural comedians of our time, and eventually brought country humor recognition and respect on a national level.

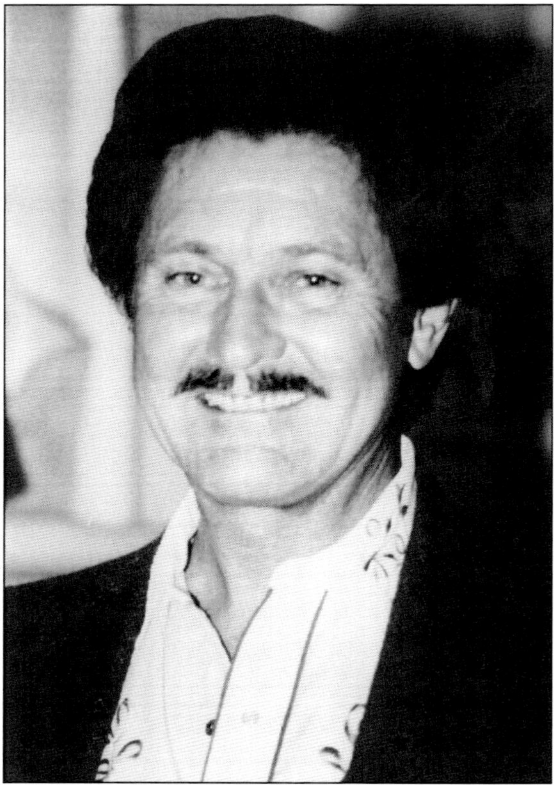

New acts, including Johnny Shelton (left), played regularly on the *Barn Dance* at WNOX, and Shelton did so again when the show was revived in 1982. He went on to make a name for himself as a local musician. Singer Mary Mincey (below) worked for WNOX and performed on stage at the *Barn Dance* in the 1980s. (Both courtesy of Edith Kirkland.)

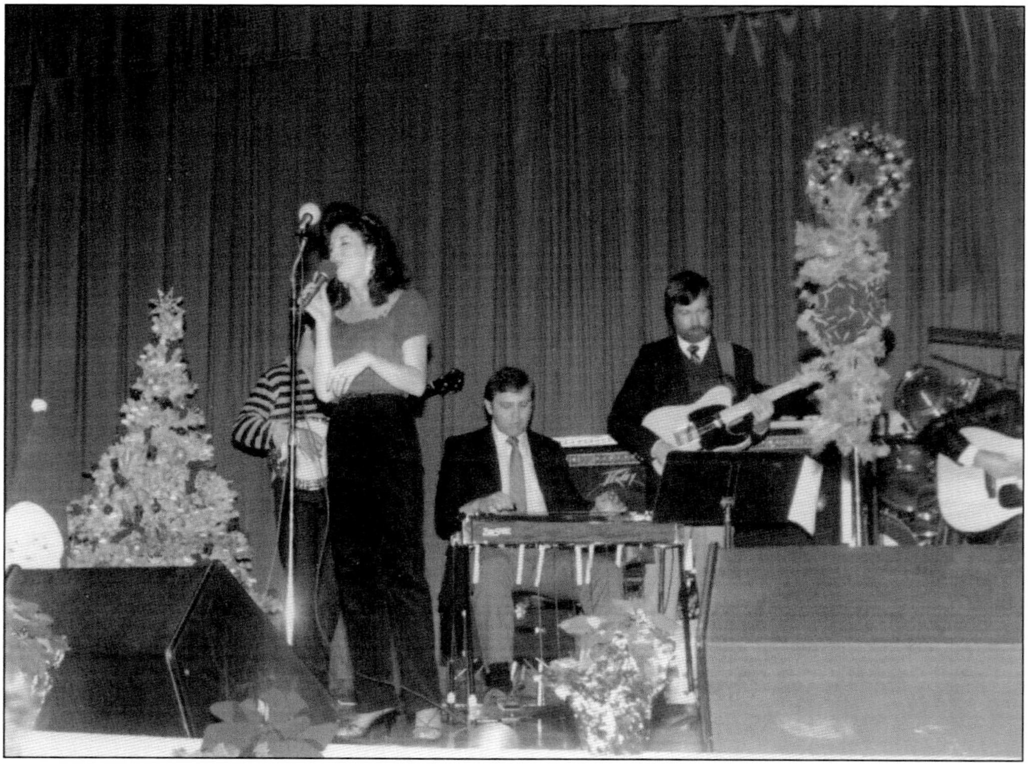

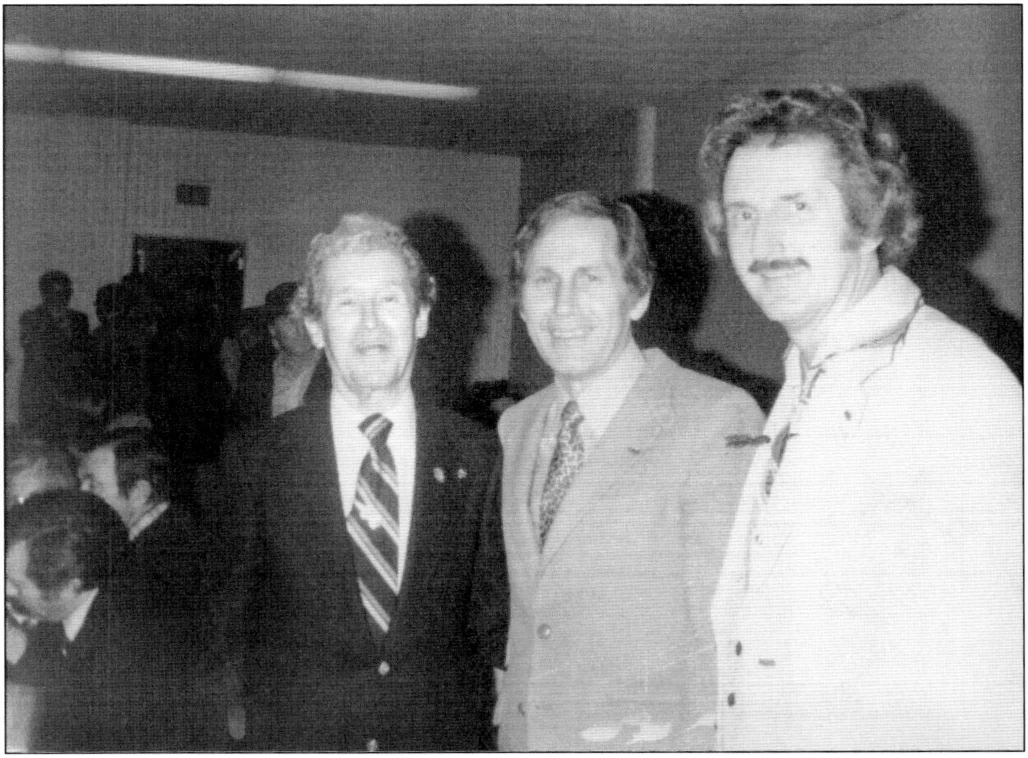

The "new" *Tennessee Barn Dance* tried to mix old performers with the new ones. Pictured above from left to right are Country Music Hall of Famers and former WNOX players Roy Acuff, Chet Atkins, and gospel singer Glenn Farrington. Below is a photograph of one of the original *Mid-Day Merry-Go-Round* players, Homer Harris, who is known for the song "Little Brown Jug," which won him a radio contest in Oklahoma and was featured in a Hollywood film. He was known as the "seven-foot-tall smiling cowboy" to many Western fans who followed his career in film and on stage. (Both courtesy of John Hitch.)

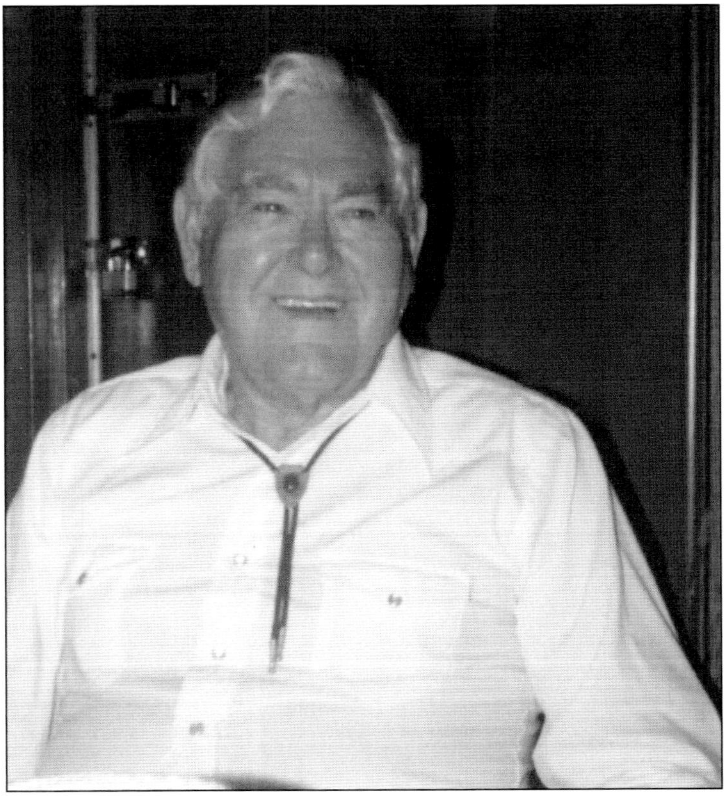

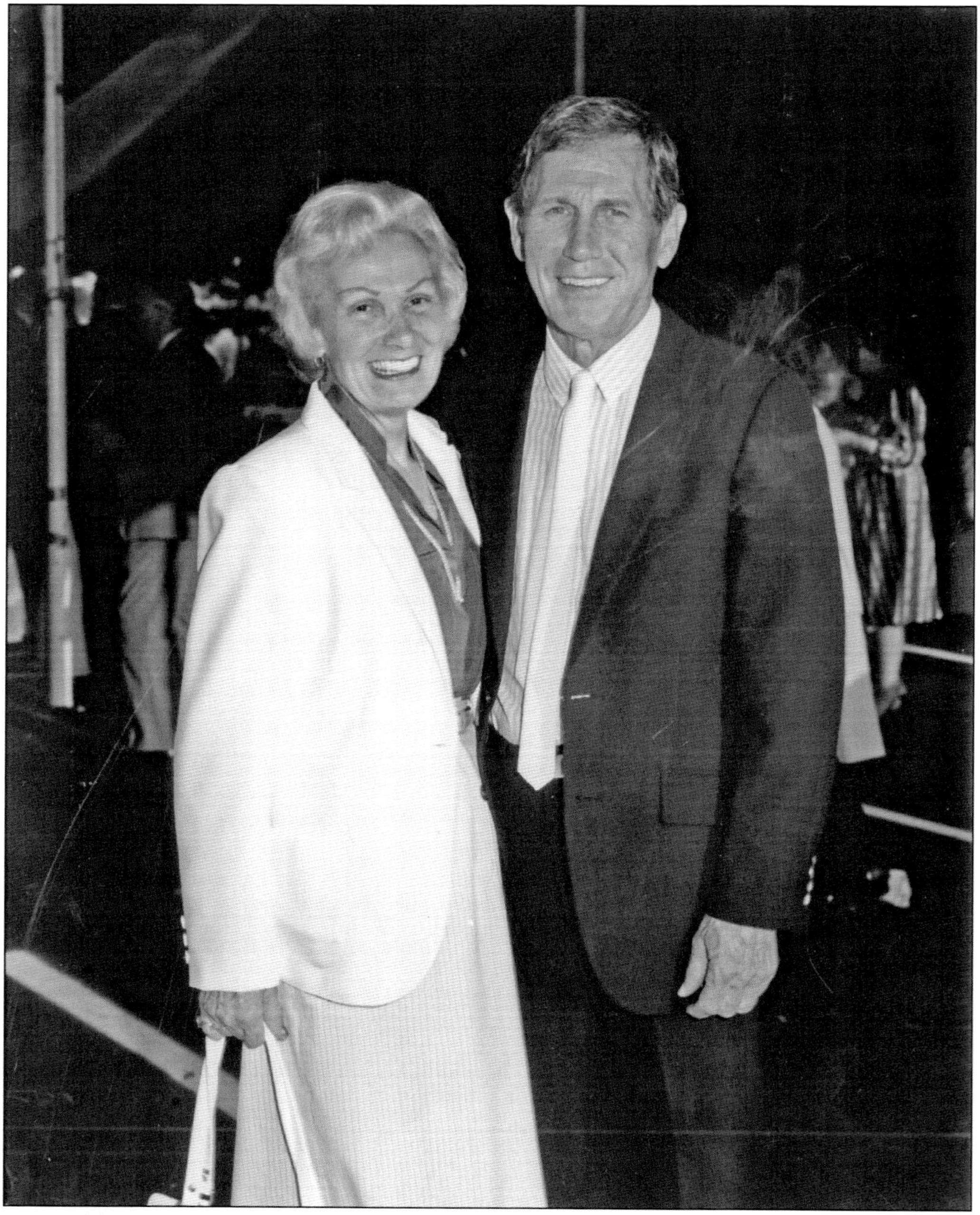

Pictured here is Country Music Hall of Fame guitarist Chet Atkins and his wife, Leona, who took time to pose for the camera at the 1985 *Tennessee Barn Dance* dedication ceremony. Chet Atkins is one of the most recognized names in Knoxville from WNOX. He and Lowell Blanchard remained friends throughout their careers, and Atkins never forgot the man who helped give him his first start in the music business. (Courtesy of John Hitch.)

Five

WNOX Goes Top 40

In the 1950s, rock and roll was introduced to popular culture. It was a new sound but still distinctly American. The 1960s brought the Beatles and the British invasion. Rock-and-roll record sales boomed, and the new music format changed the sound of radio everywhere.

Dramas, soap operas, and variety shows continued on radio but started falling out of favor with listeners as television became more popular. The aging demographic listening to old-school radio was not something station owners considered attractive to advertisers, and many stations started moving to the new rock-and-roll format.

Scripps Howard wanted to stay ahead of the curve and made the switch to rock and roll in 1962. It was still an Scripps Howard station, however, and the new format did not change the rules of how radio was done, nor did it lessen its service contracts with the community. Good broadcasting was good broadcasting, and the working philosophy was the station gave listeners what they wanted to hear in music while still providing what they needed to hear in news and information. They still staffed news and sports reporters, and announcers were still required to have a Federal Communications License to work on the air. There were only three television networks, and radio retained a powerful influence.

It was a new generation for broadcaster announcers. Radio had become more of a science but still required creative minds to add the new music and maintain the old formula. Lowell Blanchard of *Mid-Day Merry-Go-Round* and *Tennessee Barn Dance* fame had found new life as a sports announcer for the station and was working with the semi-professional baseball team the Knoxville Smokies and the University of Tennessee Athletic Department. WNOX was still the official Emergency Broadcast Station of East Tennessee and carried the UT basketball games.

WNOX found itself in another pioneering situation in 1962. Rock and roll was not exactly popular in the "polite" sections of the community any more than country music was when WNOX starting pouring it over airwaves. They brought the music into Knoxville with a more subtle approach than they had the old "hillbilly" bands. The biggest difference in the music formats for announcers was the constant movement of keeping up with popular culture. One-hit wonders were more prevalent in rock and roll, and young listeners were not as loyal to performers as were country fans. What was a fad one week was discarded the next, and announcers had to stay on their toes. It took a unique knowledge of the music to find a format that worked well and WNOX did just that.

When Scripps Howard made a commitment, they went all out, and they did with the new rock-and-roll format. For the next 20 years, the radio station reigned as the one to beat. The competition was fierce with WKGN and other stations that would rise and fall on the ratings. There was still enough of the old to blend well with the new.

The 1960s and 1970s were a roller coaster ride, and the Top 40 Charts reflected a constant sea of change of musical tastes in the field of rock and roll; however, early on, WNOX found a music director in a former television personality who devised one of the most unique music format formulas to keep and hold the highest amount of listeners. When the new music hit the airwaves, it was not really new at all. Rock and roll was as regional in the United States as country and western music was when it began. This music director came up with a plan to merge them all into one unique sound for WNOX. It gave the station an edge others did not have with the new format and helped polish the erratic image of rock and roll among industry professionals as well as listeners.

WNOX's station colors, public service commitment, and economic presence in the community remained unchanged. The younger generation was listening, but it was a "grown up" radio station entering the second age of radio broadcasting.

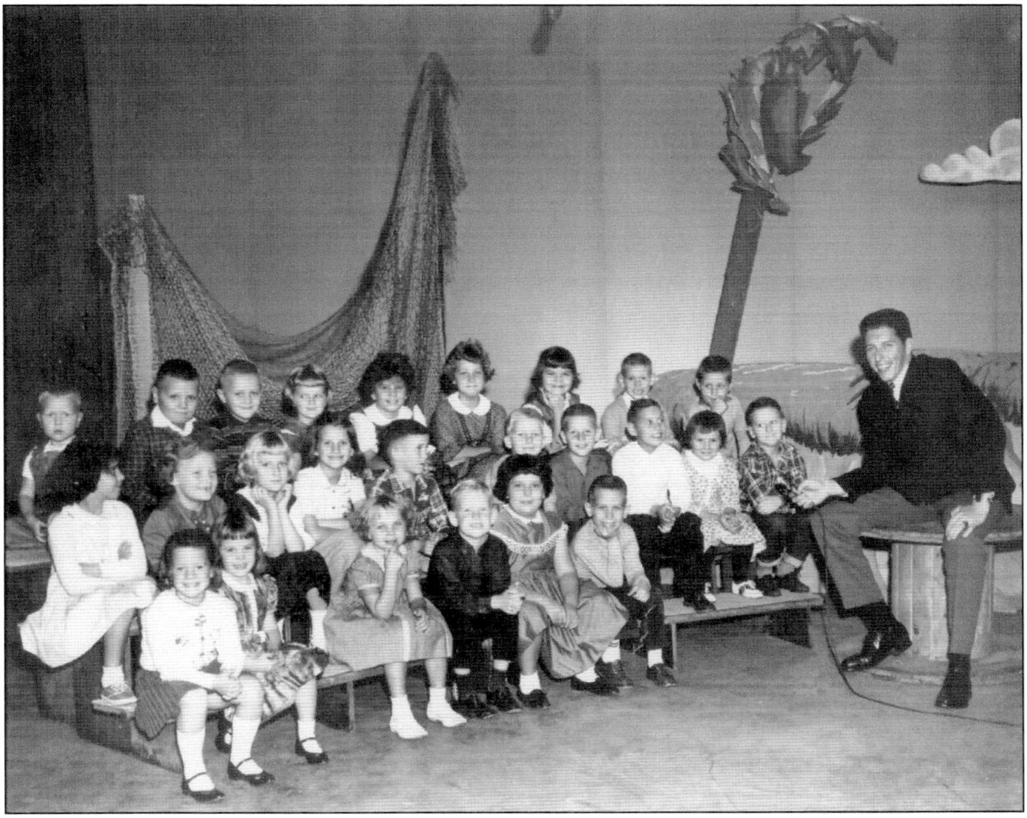

John Pirkle started his broadcast career at WATE-TV. He is pictured here in 1965 hosting the *Popeye Show* for local children. The daily program was one of Knoxville's most popular children's programs. He left the job to take a position at WNOX in 1966. (Courtesy of John Pirkle.)

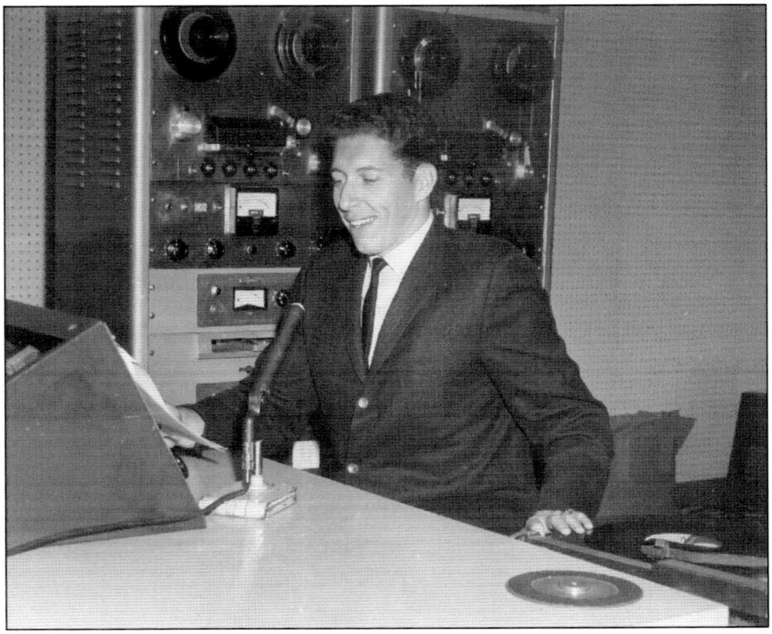

Pirkle was the afternoon drive announcer for the station and is seen here at the console of WNOX pulling an air shift. He was known as a hard worker at the station and began studying a way to blend and mix the new music that would make it appeal to the widest number of listeners. (Courtesy of John Pirkle.)

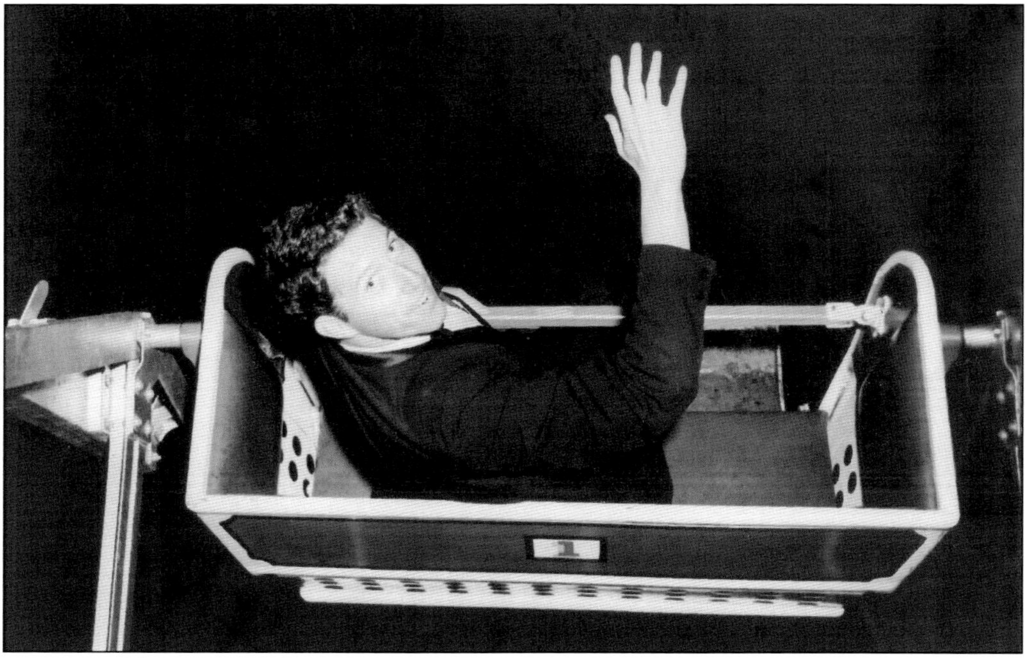

In 1964, Pirkle spent a week on a Ferris wheel at the Tennessee Valley Agricultural and Industrial Fair, where he logged 127 hours riding it. The promotion was so unique it was picked up as a story by Associated Press and United Press International. Below, his then-girlfriend and now wife of 42 years, Manya, joined him for the last hours of the event. (Both courtesy of John Pirkle.)

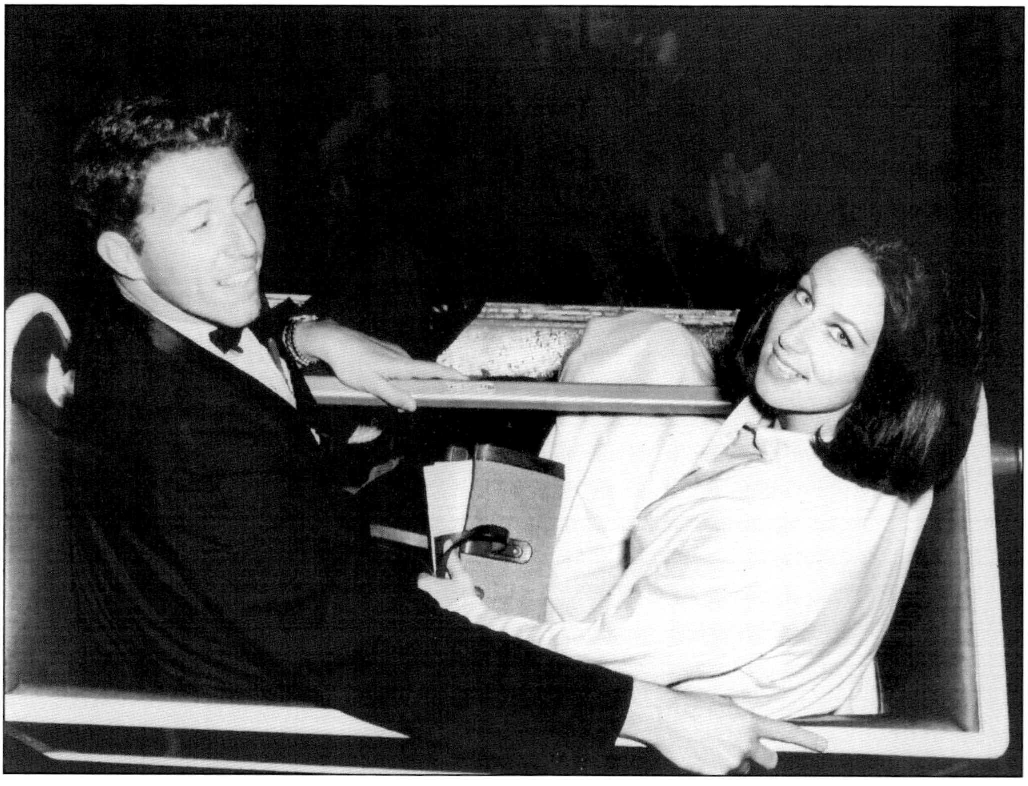

Here Pirkle is surrounded by listeners and well-wishers, who had been following his Ferris wheel ride at the TVA&I Fair on the radio. The promotion raised WNOX's ratings, and the station went on to become the number one station for years. (Courtesy of John Pirkle.)

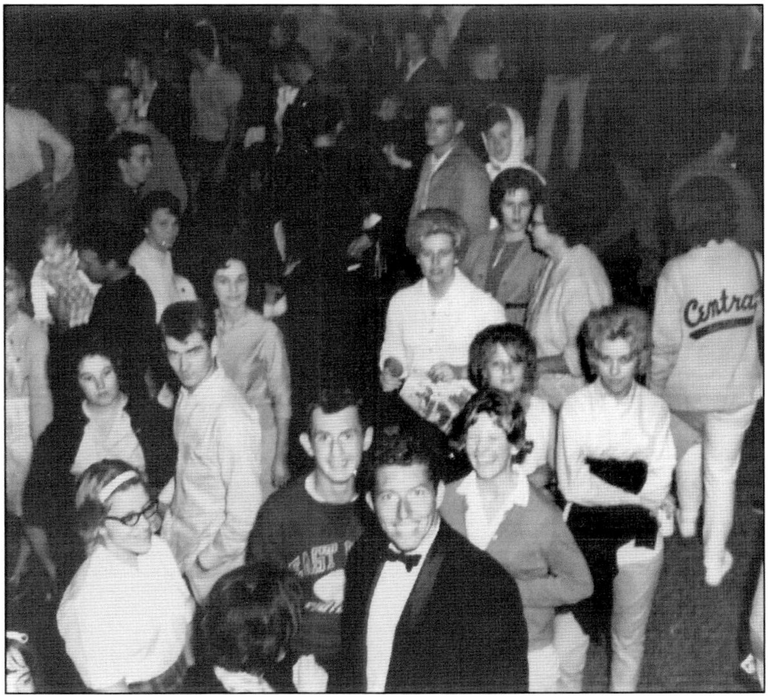

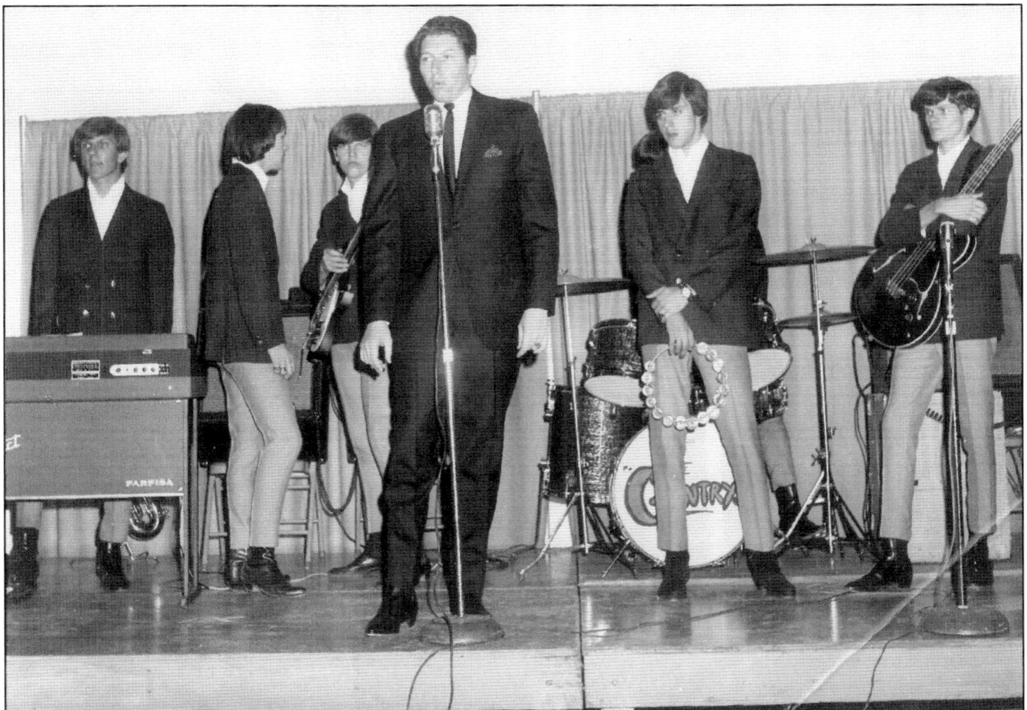

Pirkle was also a talent manager on the side and hosted bands throughout Knoxville. In addition, he was a talented master of ceremonies. Pictured here is the Memphis band the Gentrys. They were one of many recording groups WNOX brought to Knoxville. The band was known in the 1960s for their number four single "Keep on Dancing." (Courtesy of John Pirkle.)

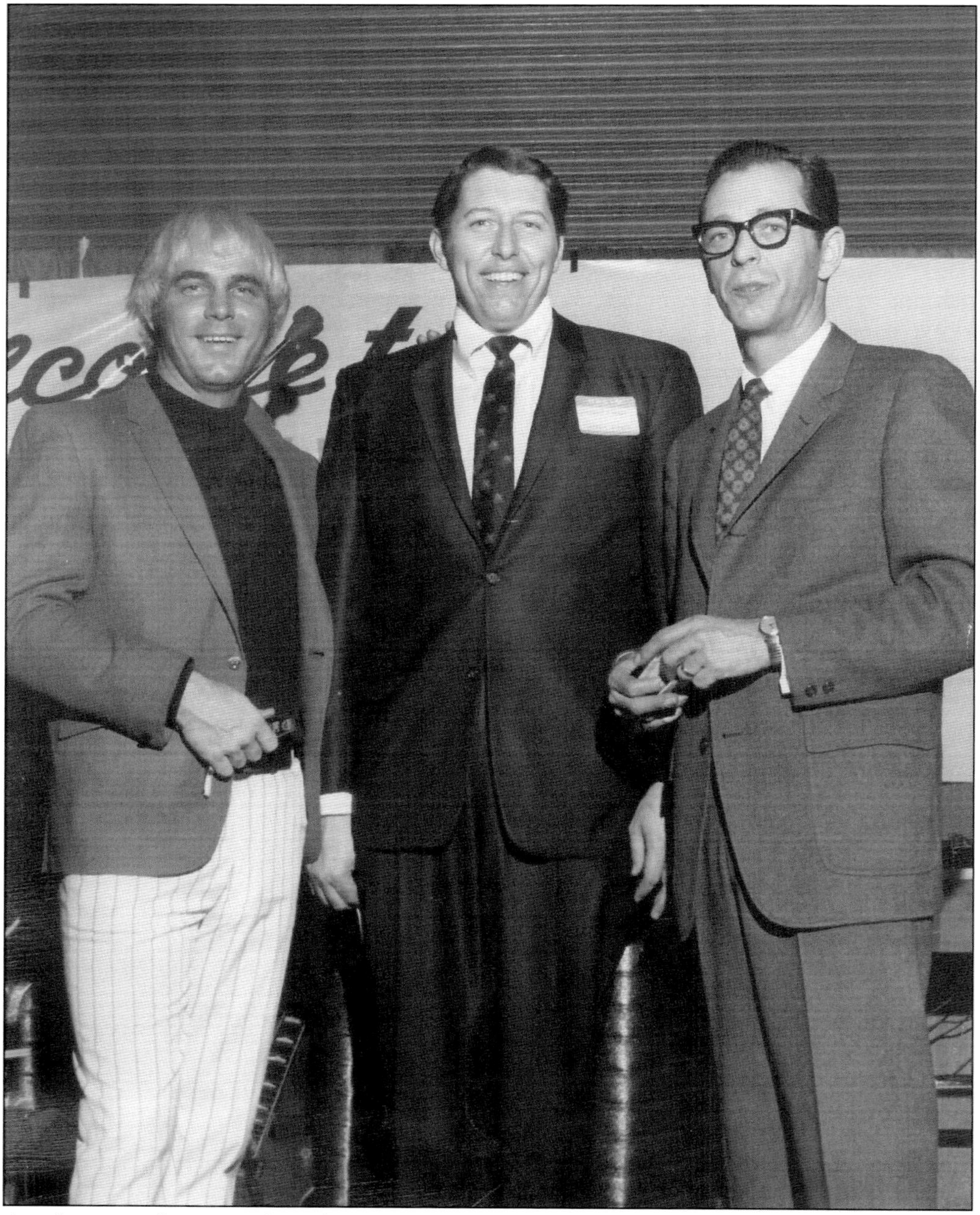

Another recording artist brought to Knoxville by WNOX was Joe South, who was known for hits such as "Birds of a Feather" and for winning three 1969 Grammy awards, including best song of the year for "Games People Play." He was a songwriter who also penned the hits "Down in the Boondocks" for Billy Joe Royal and "I Never Promised You a Rose Garden," made famous by Lynn Anderson. From left to right are Joe South, music director John Pirkle, and WNOX general manager Pete Dreyer. (Courtesy of John Pirkle.)

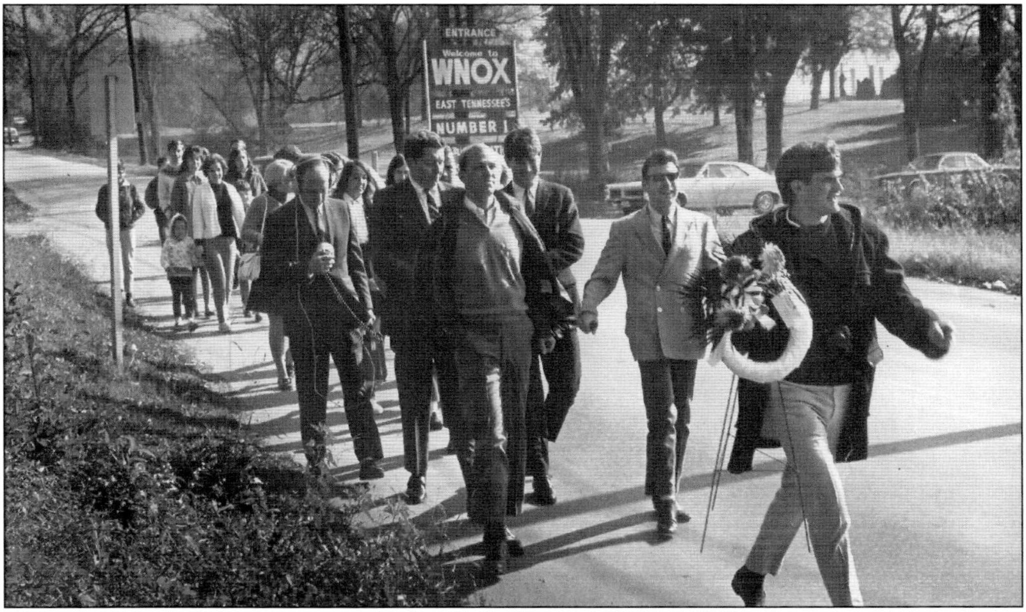

In 1968, Pirkle came across a 45-rpm in a stack left by a record representative. He played it and liked the tune, so he got a marker and crossed out the artist and song title where no one could read it and stuck it in rotation to be played every hour. WNOX became the first station in the nation to debut the song "Ode to Billy Joe" by Bobbie Gentry. By the time other stations learned the title of what was the hottest record of the year, Pirkle and crew made a big deal of already wearing the song out on the air and held a mock funeral for it. This shows the station personnel in a funeral procession. It began at the WNOX studios and was a march down Broadway. Below, cars pulled over and stopped as a courtesy, which is common for funeral processions in East Tennessee, and allowed the group to pass solemnly. (Both courtesy of John Pirkle.)

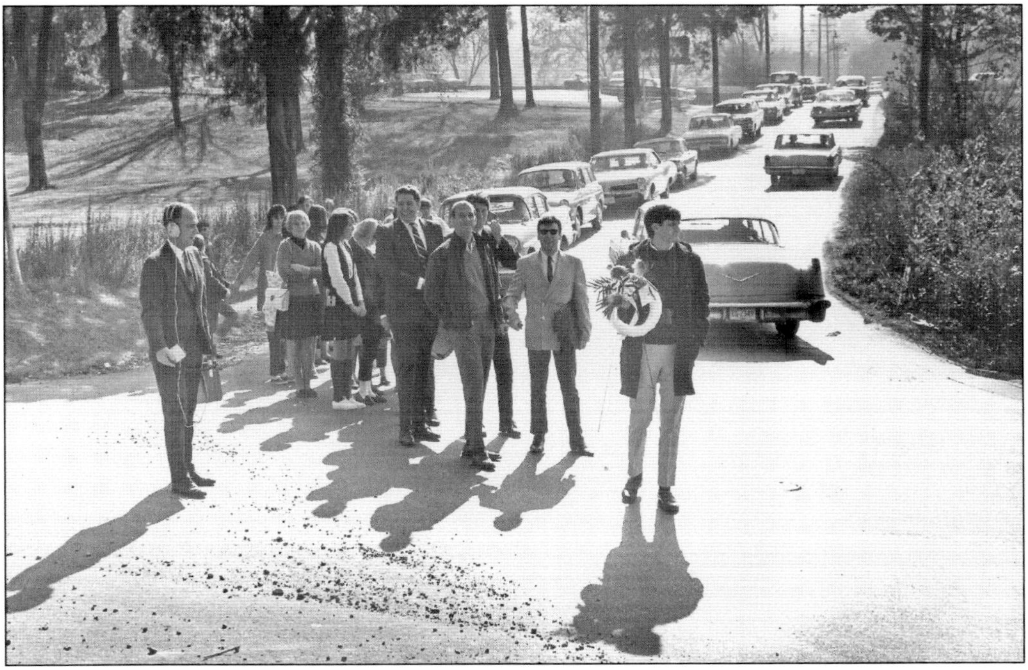

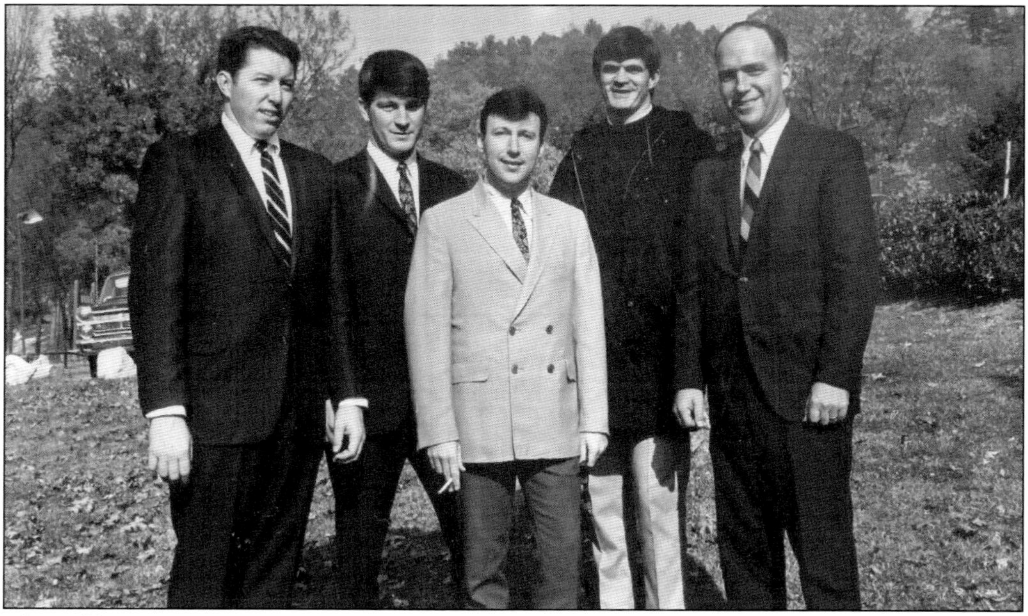

The pallbearers for the funeral were, from left to right, John Pirkle (music director); announcers Phillip Conn, George McGovern, and Rob Galbraith; and program director Jim Clark. Galbraith would move on to Nashville to work as a producer for Grammy Award–winning country entertainer Ronnie Milsap. Below, the station staff and passersby who happened on to it stood in silence as a funeral was held for the song "Ode to Billy Joe." The news department, armed with field equipment, covered the event for later playback to listeners. (Both courtesy of John Pirkle.)

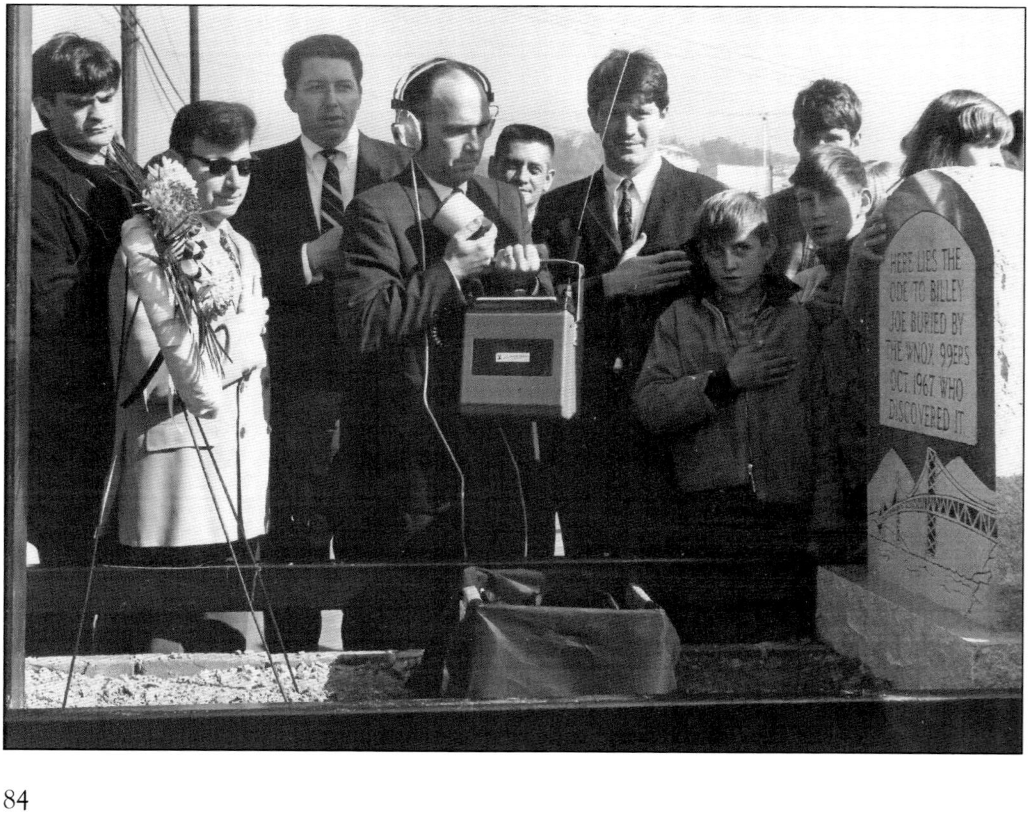

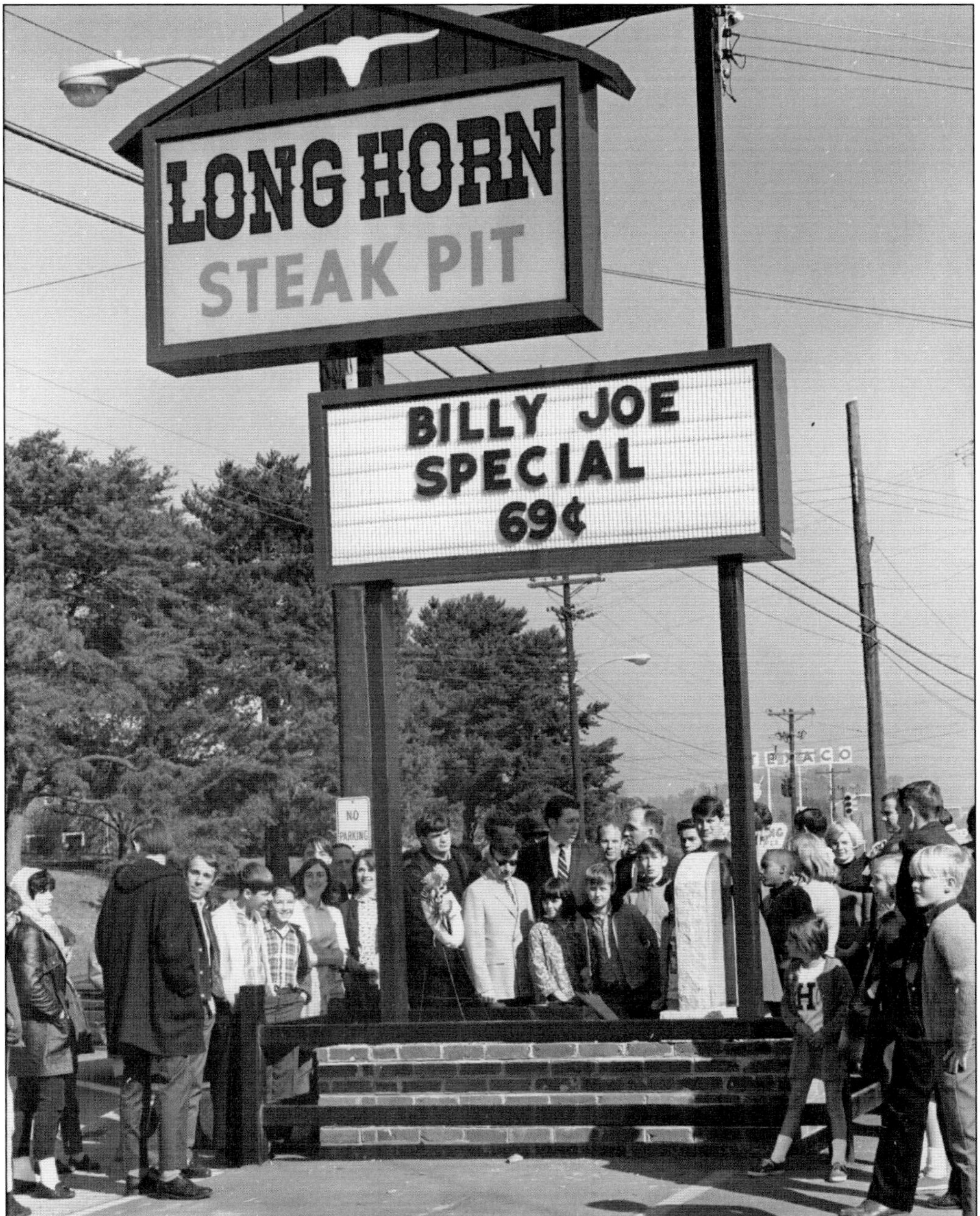

Following the funeral, the station staff, mourners, and the curious who stood by in respect broke for lunch at Longhorn Steak Pit, where the Billy Joe Special was 69¢. It was the last time "Ode to Billy Joe" was heard on WNOX. (Courtesy of John Pirkle.)

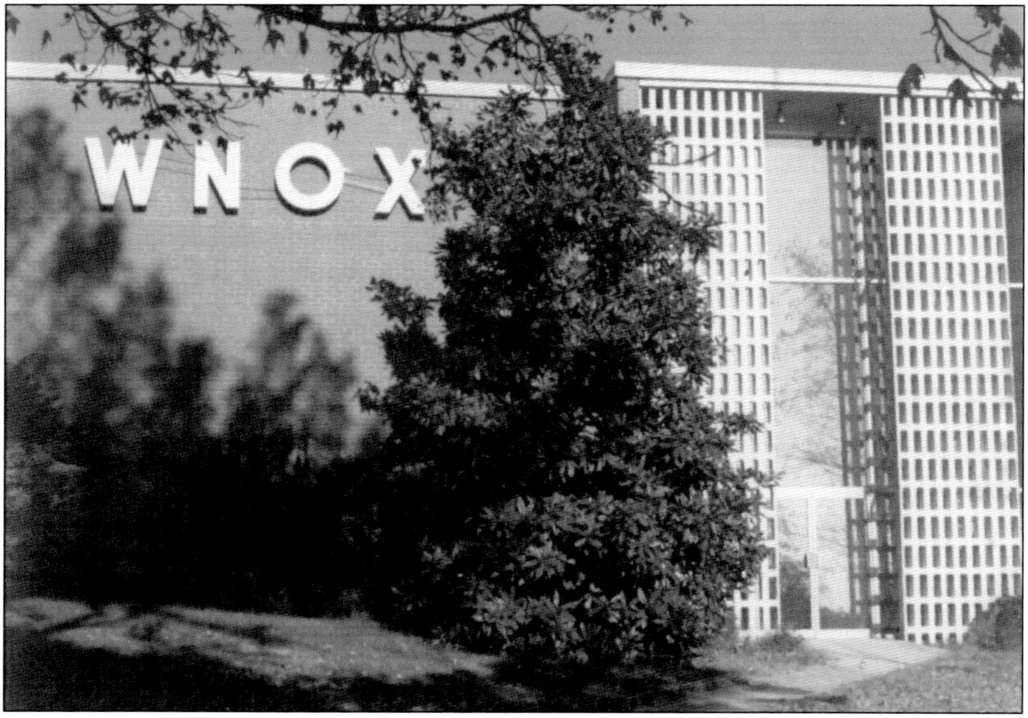

The above photograph was taken of the outside of the WNOX studios in the 1970s. There were other competitors in the market. WKGN and WRJZ each went full press on promotions and station awareness campaigns to attract listeners. The station, with its unique musical format, was entrenched by then as one of the leading rock-and-roll stations with four of the most recognizable call letters in the United States, and as in the early days of the organization, record label executives sought it out to introduce new artists.

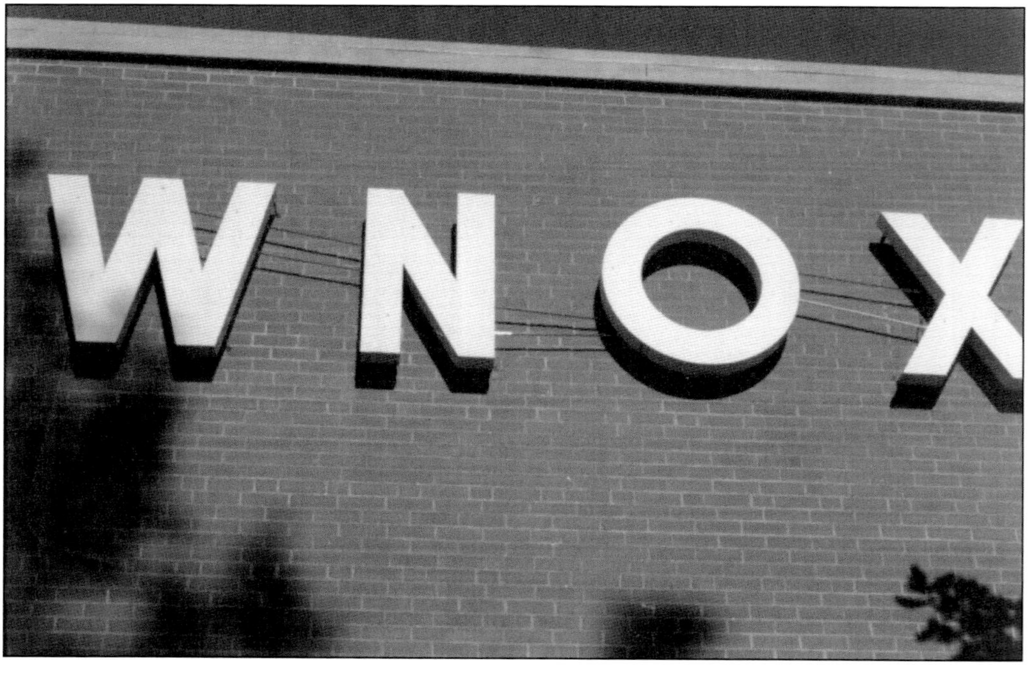

John Pirkle is pictured here on the hotline in the control room. As music director, he came up with the unique formula of playing 50 percent Top 40 and 50 percent rhythm and blues, with 60 percent of the R&B being the Memphis and Philadelphia sound and 40 percent Motown. The blend of the music gained respect in the industry for pulling in a maximum number of listeners with a unique format. (Courtesy of John Pirkle.)

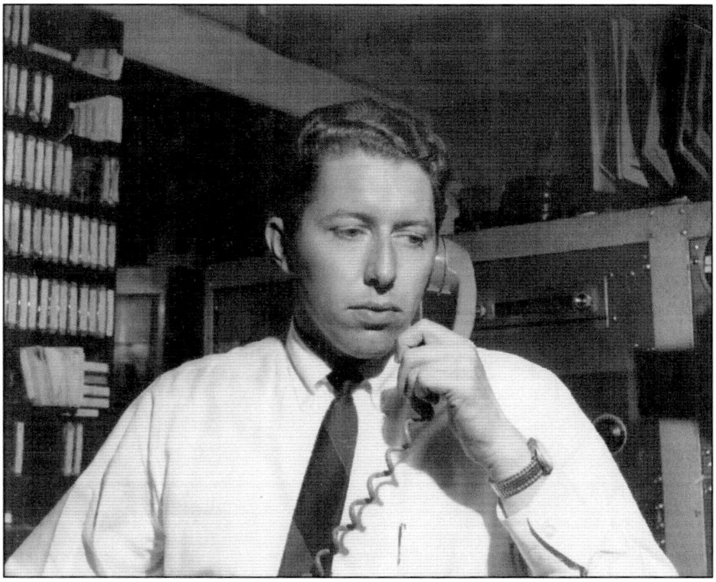

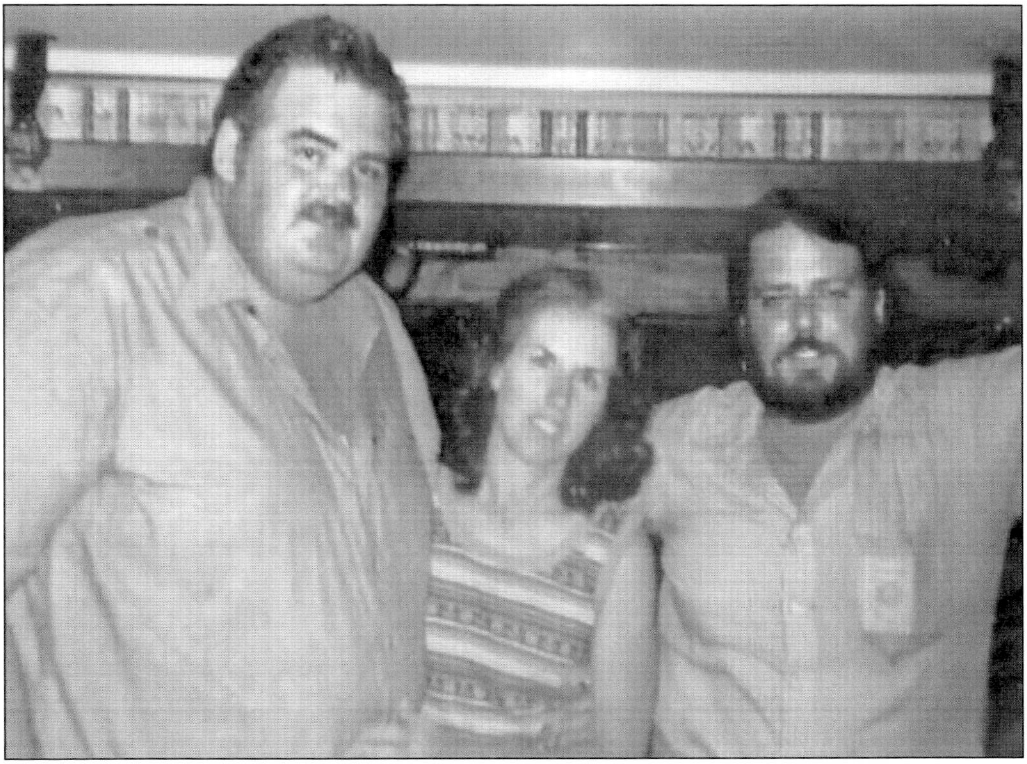

From left to right are WNOX program director Rex Miller; his wife, Carol; and John Long. Following Otis Redding's 1967 concert in Knoxville, Long ("Kincaid" on the air) and Rex Miller stopped in to see Eddie Beacon, who was on the air at the transmitter. Fierce rival WKGN's program director Joe Sullivan called Long; when told who it was as he was picking up the receiver, Long uttered "Joe Sullivan sucks." A listener in North Carolina was recording the show when the comment accidentally bled through on the broadcast. It was exhibit one in the lawsuit brought by Sullivan against Scripps Howard and cost all three WNOX men their jobs. (Courtesy of John Long.)

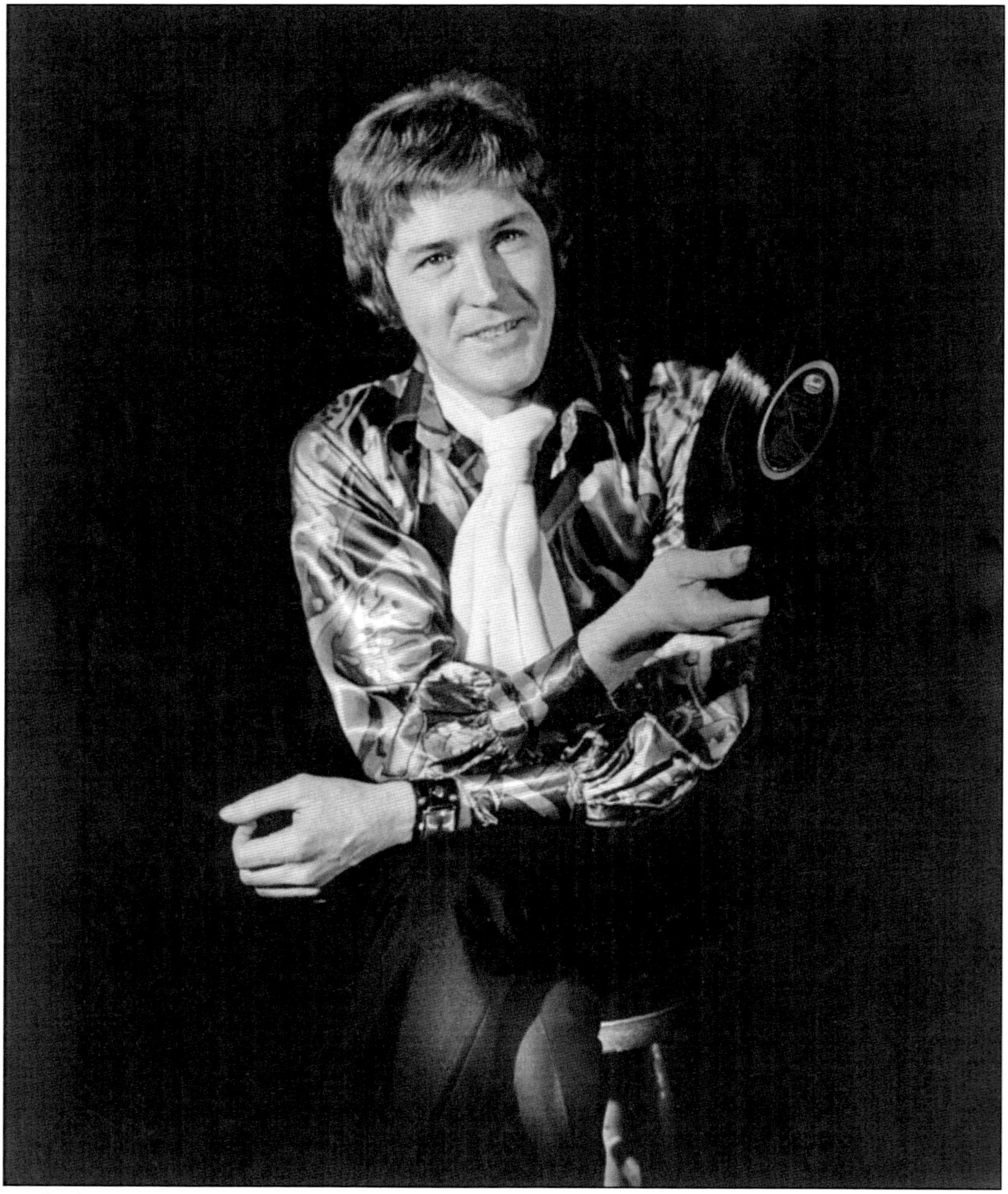

With the British invasion into rock and roll by the Beatles, the Rolling Stones, and other UK bands in the 1960s, WNOX general manager Pete Dreyer hired an announcer he named "Sir Bernard" (Bernie Quayle) in Great Britain and began a massive media promotion on the air about the "Knight of the Turntable" hired by the station. This is the publicity photograph used by WNOX to promote the announcer. (Courtesy of Bernie Quayle.)

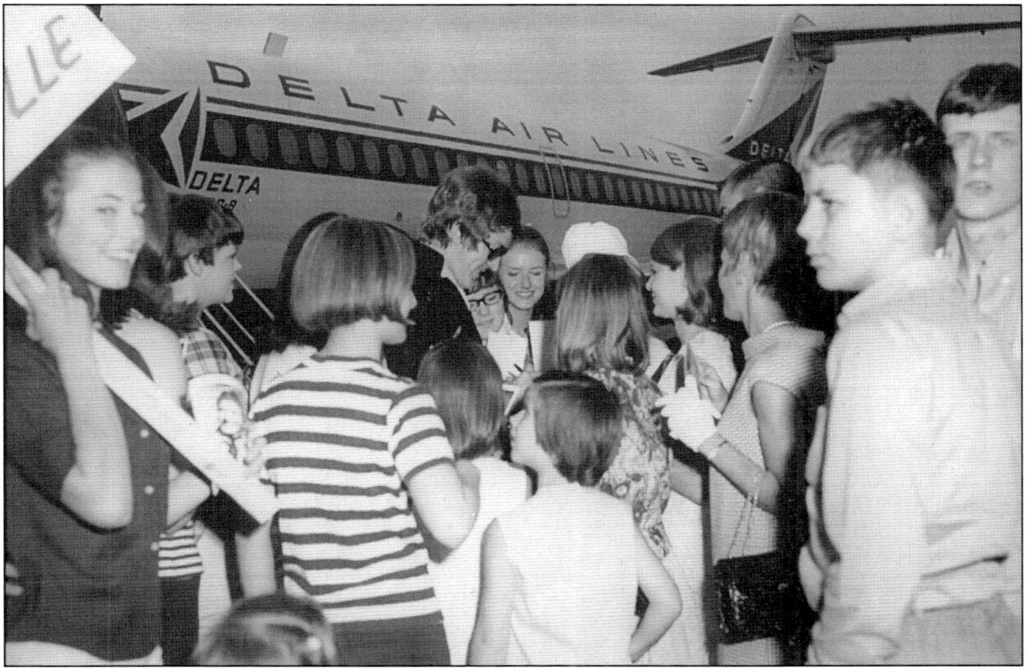

When the Delta Airlines jet touched down at McGhee-Tyson Airport in Knoxville, the young British DJ was surprised by the crowd who greeted him on the tarmac. He was told to remain until the other passengers deplaned, and a red carpet was literally rolled out for him. (Courtesy of Bernie Quayle.)

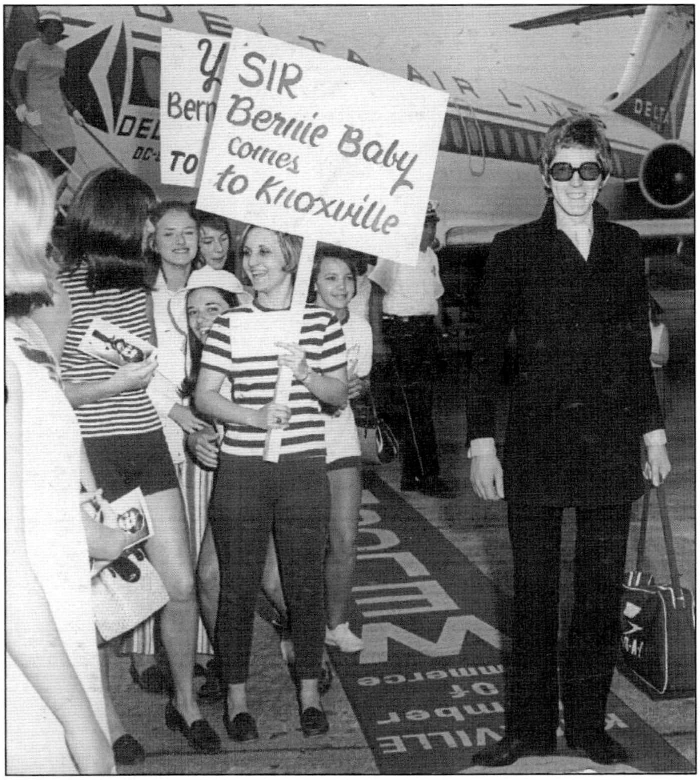

It is estimated that 2,000 people were on the runway to greet him, waving banners with his name on them. Pete Dreyer and program director Dan Armstrong, in true WNOX fashion, had made Quayle a household name in East Tennessee. He was greeted by the UT Marching Band playing "God Save the Queen" and presented a key to the city, and a small aircraft was hired to take off with a banner saying "Welcome to Knoxville Sir Bernard." (Courtesy of Bernie Quayle.)

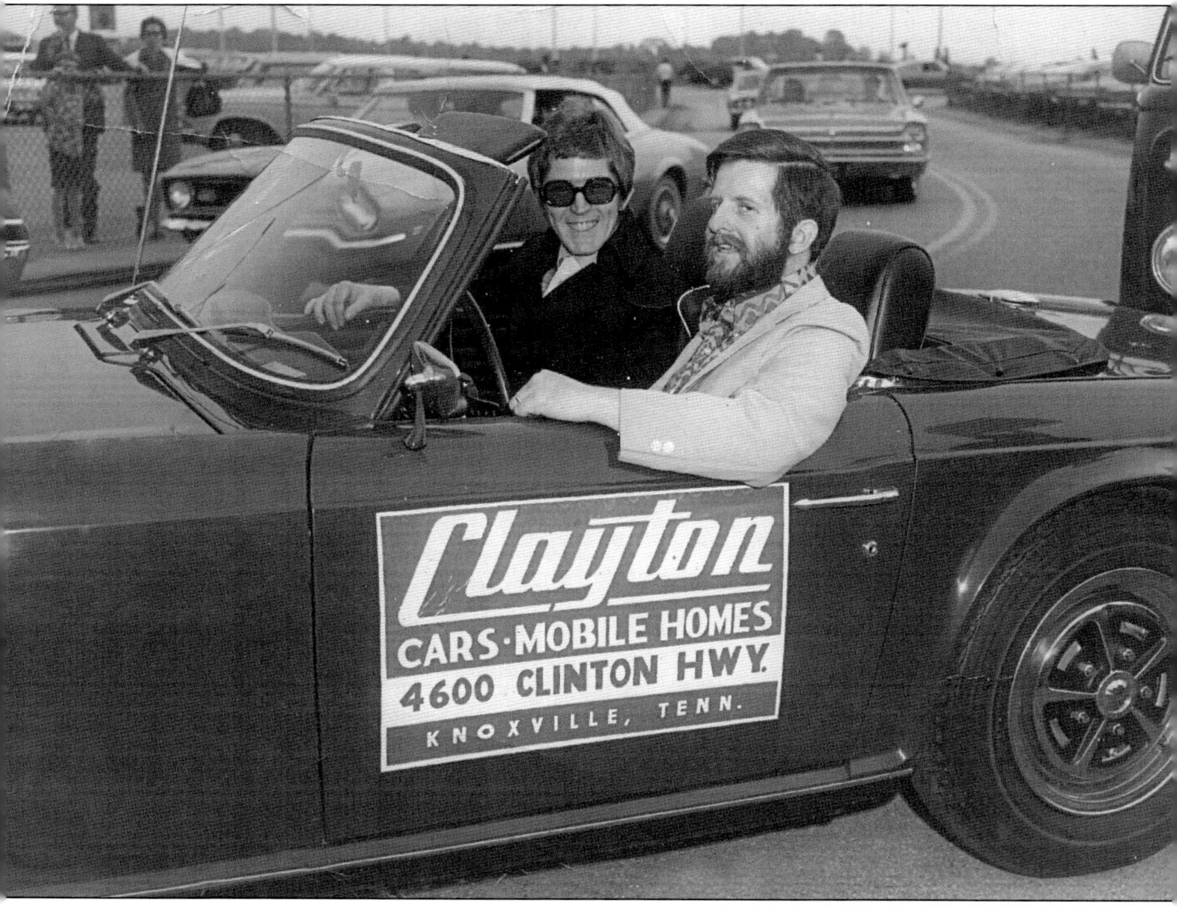

Pictured here is WNOX program director Don Armstrong at the wheel of a Triumph TR6 with Sir Bernard. It was the first Triumph TR6 in the United States and was on a six-month loan to Bernie Quayle. The vehicle was later painted with Quayle's name and the WNOX call letters. (Courtesy of Bernie Quayle.)

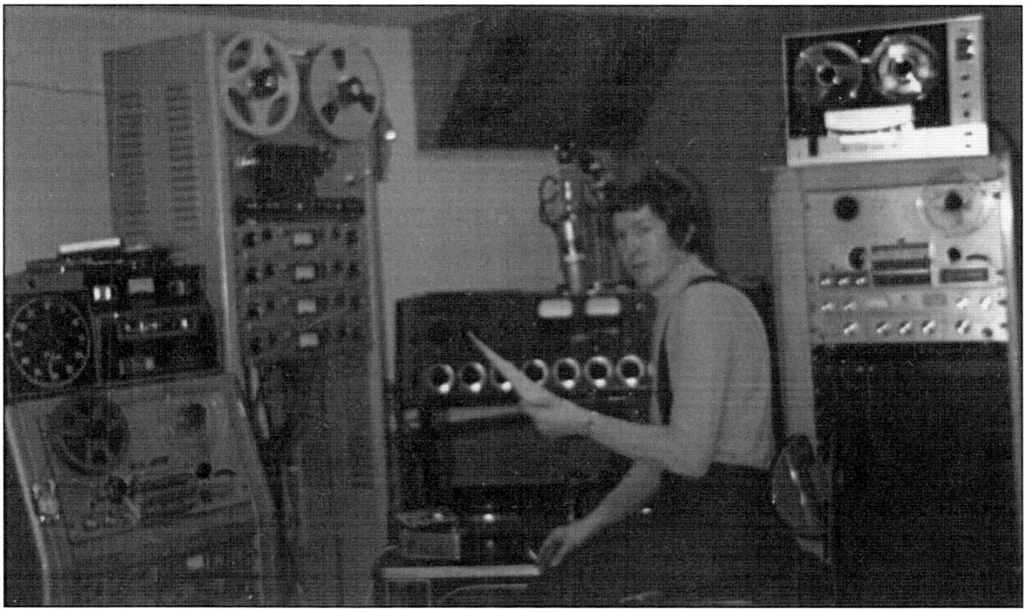

The story of Sir Bernard swept across the nation, and WNOX had a real celebrity on their hands. It almost came apart for the station when, days after arriving, Quayle got his draft notice. WNOX general manager Pete Dreyer was immediately on the phone to inform the draft board he was a British citizen. It was rescinded almost immediately. Bernard is pictured above in the WNOX production room. Below, Quayle poses for a photograph on the trademark WNOX circular staircase. The British announcer was surprised by national magazines like *Sixteen* and others talking about the new British disc jockey "sweeping the nation." (Both courtesy of Bernie Quayle.)

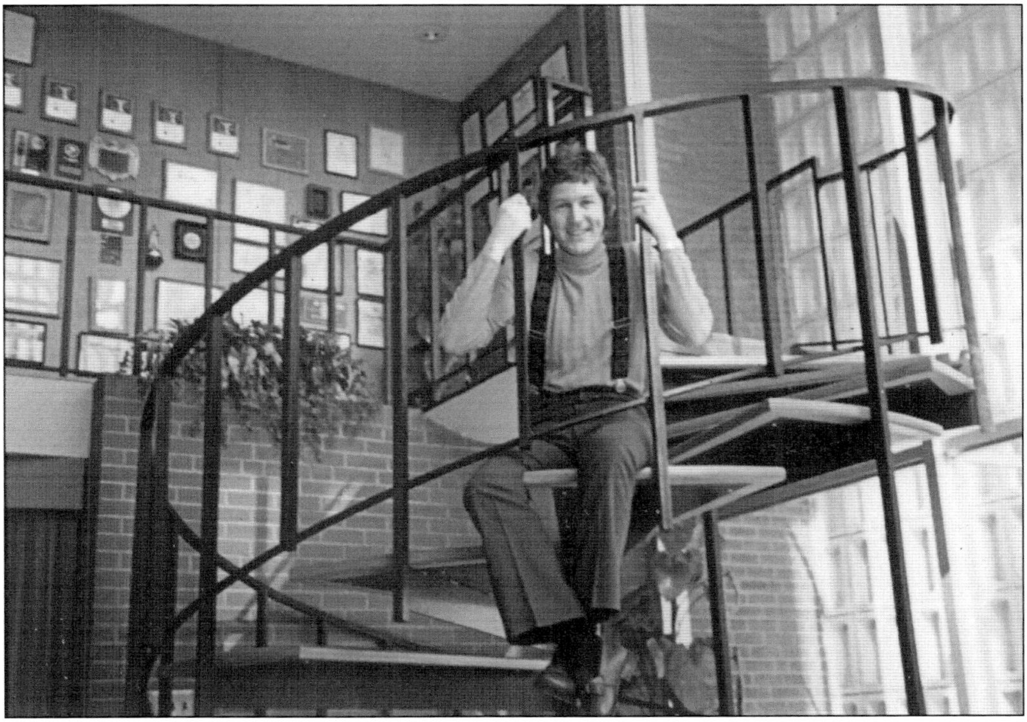

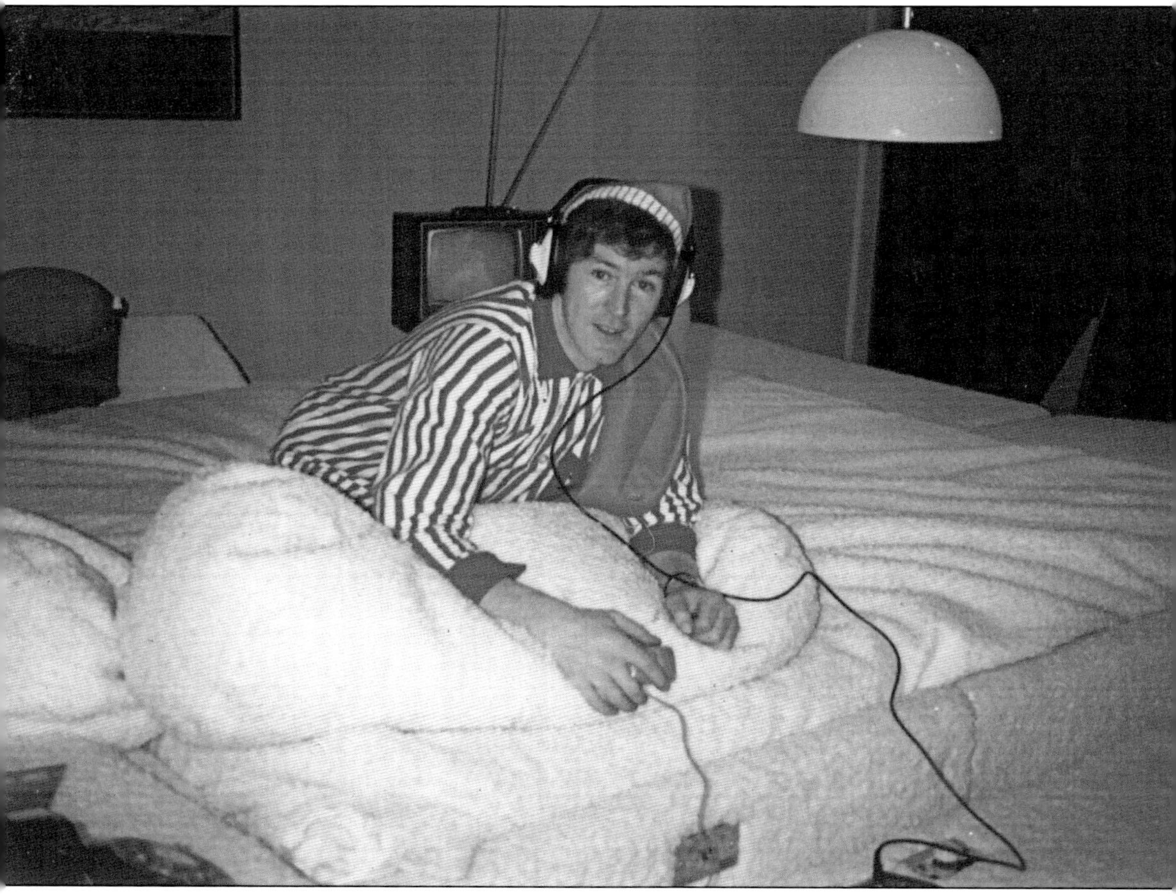

One of the big station promotions with Sir Bernard was the three days he slept on a $28,000 water bed in a furniture store on Gay Street where he did his radio broadcasts. He remembers he did not get much sleep as people kept knocking on the window and asking him to jump up and down on the bed. An intercom linkup from the street was not much help either. Quayle remained in Knoxville until 1973, when he moved to WOHO in Toledo, Ohio. He eventually returned to the United Kingdom and, in 2007, celebrated his 40th anniversary on the air at Manx Radio. (Courtesy of Bernie Quayle.)

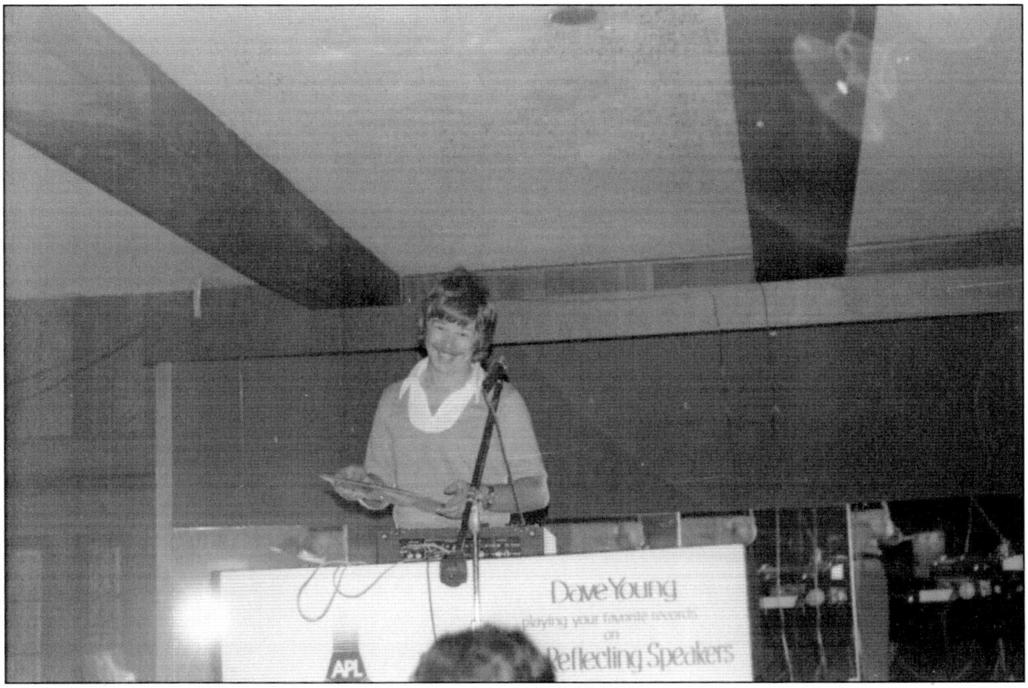

In 1968, announcer Dave Young started work at WNOX. Here he is pictured playing music for a dance. Dances were still a popular source of promotion for the radio station throughout the late 1960s and early 1970s. Announcers could pick up extra money for the work and further entrench the listeners with music. Below is a photograph taken of Young in a 1973 Opryland promotion where WNOX took station contest winners to see the popular Osmond Family in concert.

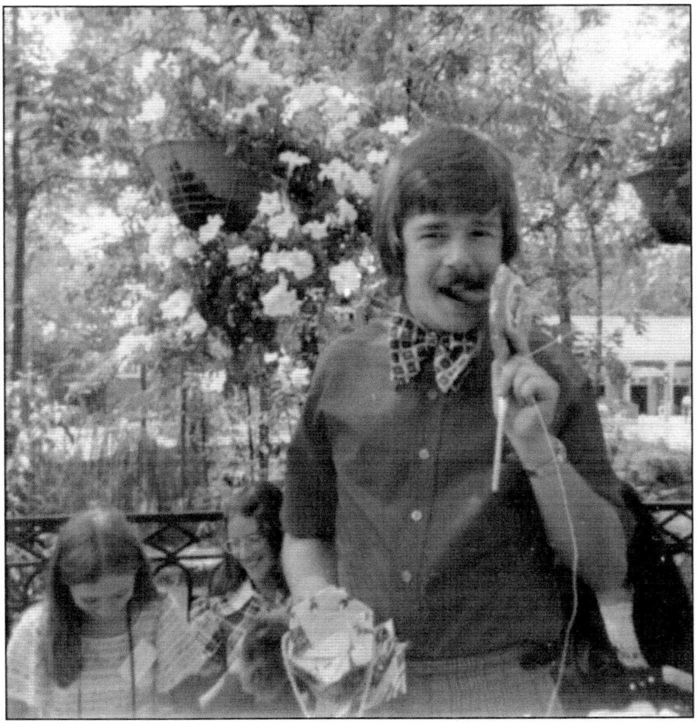

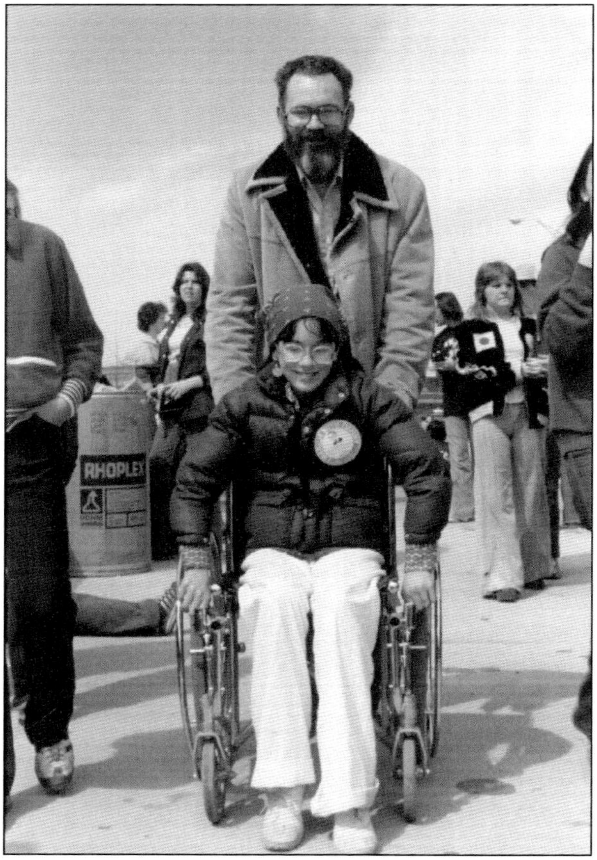

WNOX also remained a fixture in Knoxville's public service sector by holding a variety of special events aimed at raising money for charities. One of the longest running charitable fund-raisers they participated in was the annual March of Dimes, which was started at the station during the days of Lowell Blanchard. Below, station announcers take time out to sign autographs for listeners. (Both courtesy of Janice McCroskey.)

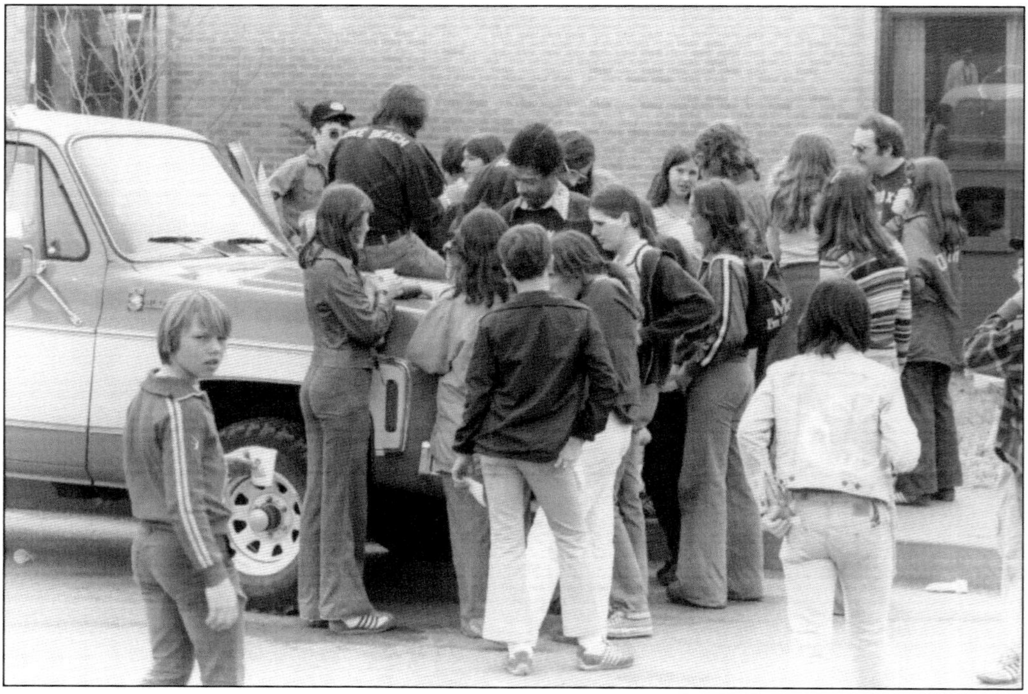

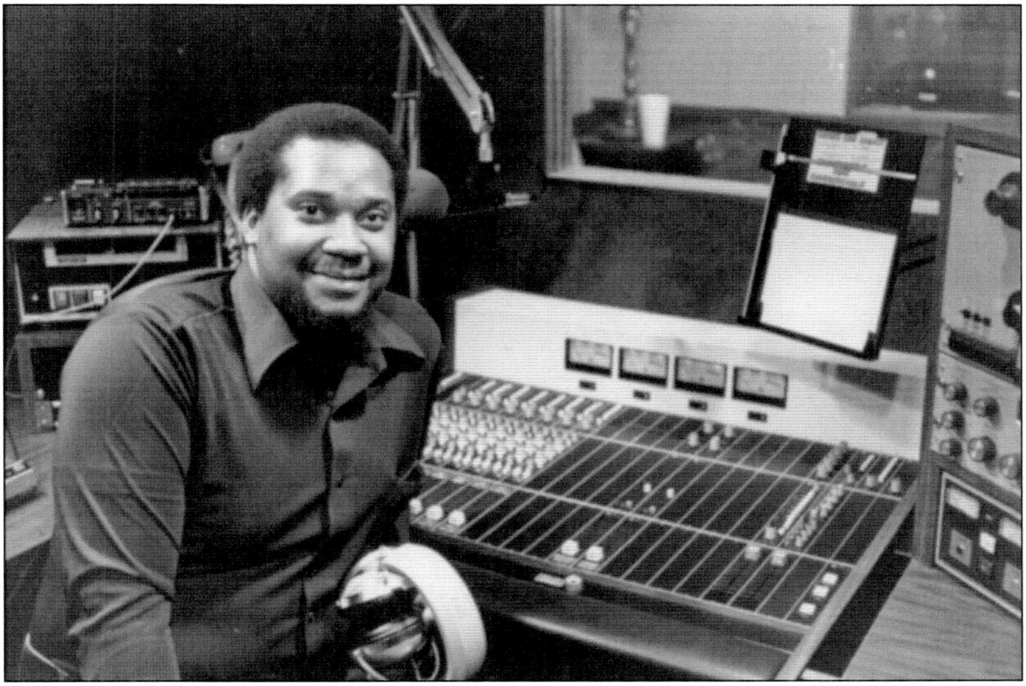

James Maurice Turk was another popular announcer at WNOX. Here he is pictured at a multichannel mixing board used by the station to record and dub commercials. At the time, WNOX was one of the leading production houses in the region for radio spot producing. (Courtesy of Janice McCroskey.)

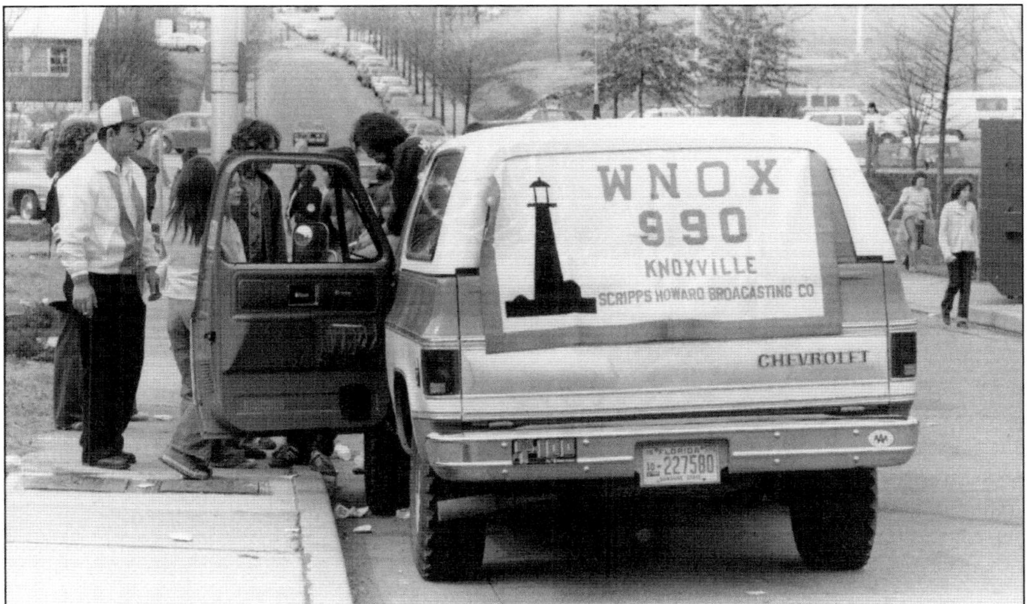

The station held live remotes at concerts to promote the station brand. Here the engineering truck is parked outside the Knoxville Civic Coliseum, which was the most popular venue in East Tennessee to hold concerts. WNOX was responsible for bringing a number of big name rock-and-roll artists to the city in the 1960s and 1970s. (Courtesy of Janice McCroskey.)

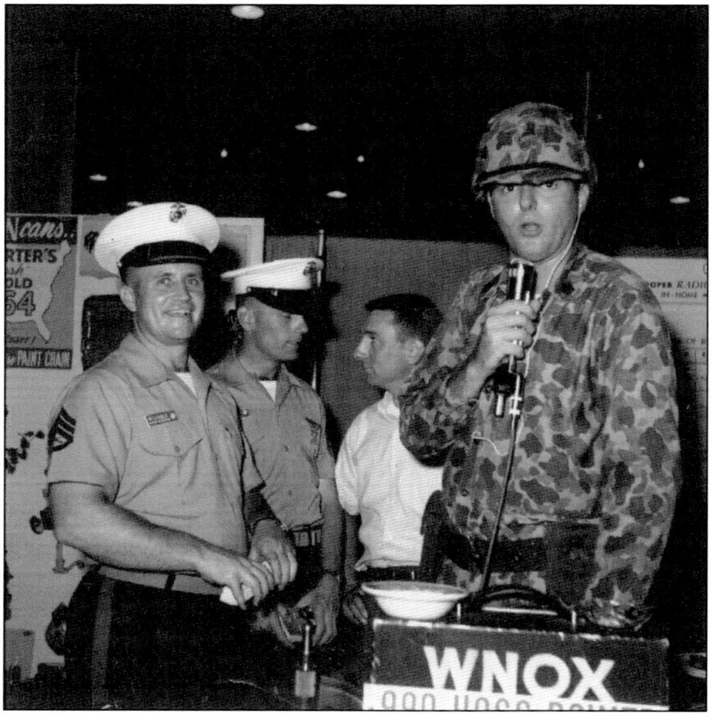

This is a 1965 photograph of a WNOX promotion with the U.S. Marine Corps. Although it was a rock-and-roll station, WNOX still reflected its listenership, which was traditionally a huge supporter of the armed forces, and supported the men and women in the U.S. military during the Vietnam War. (Courtesy of Janice McCroskey.)

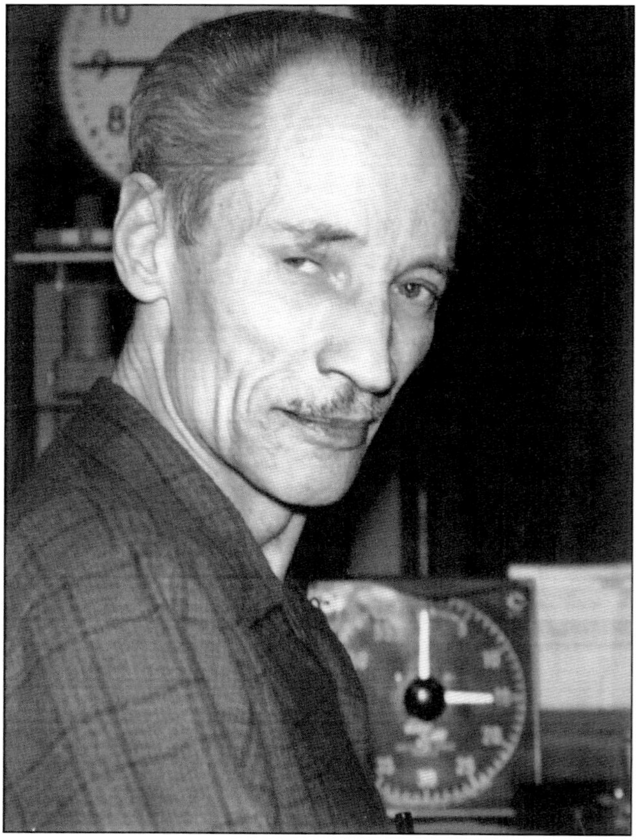

WNOX's news director and anchor Ron Ashburn was a legend for his trademark booming voice. He remained with the station throughout the 1960s and early 1970s and retired from Scripps Howard after nearly 30 years in the business. (Courtesy of Janice McCroskey.)

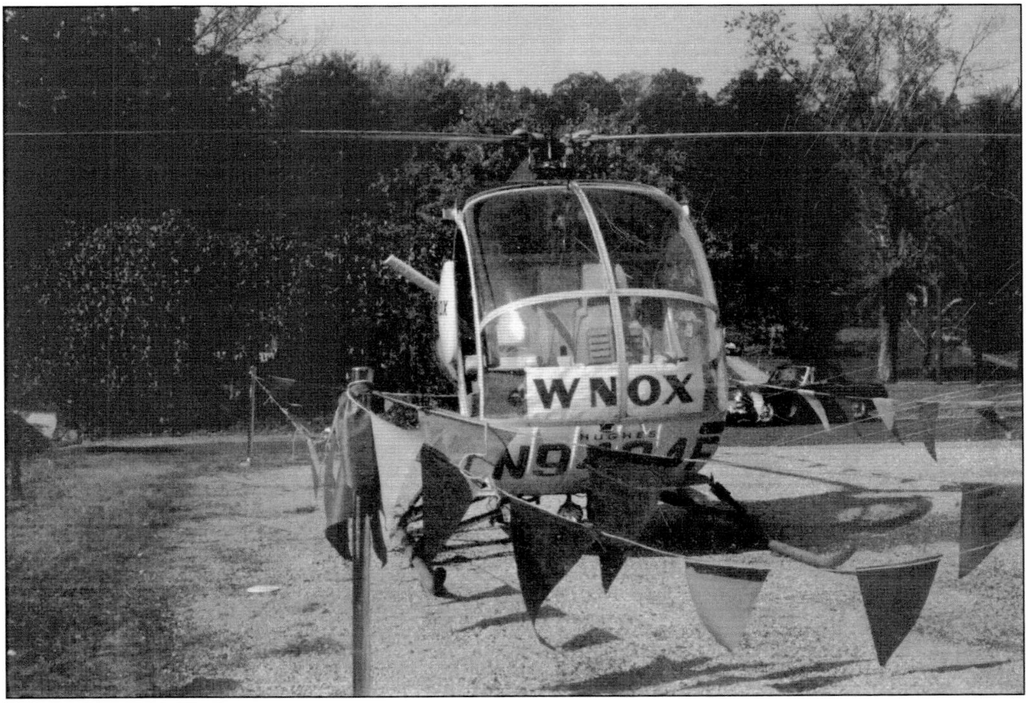

This is a photograph of WNOX's helicopter, which was used to give live traffic reports. The station was the first to implement helicopter traffic reports in the city with Knoxville Police Department officer Cletus Irwin, who was later replaced by Wayne Bell. When WNOX moved to a fixed-wing aircraft, WIVK took over the helicopter traffic reports and they became a fixture in Knoxville broadcasting. Pictured below is announcer Larry Steele (third from left) signing autographs for fans at a station event in the city. (Both courtesy of Janice McCroskey.)

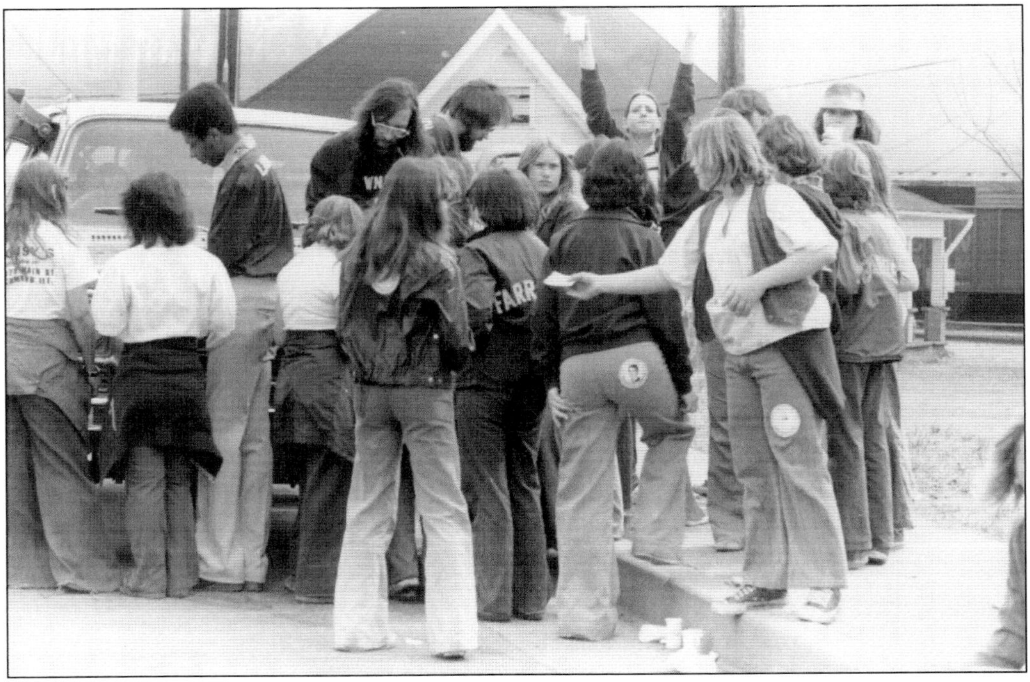

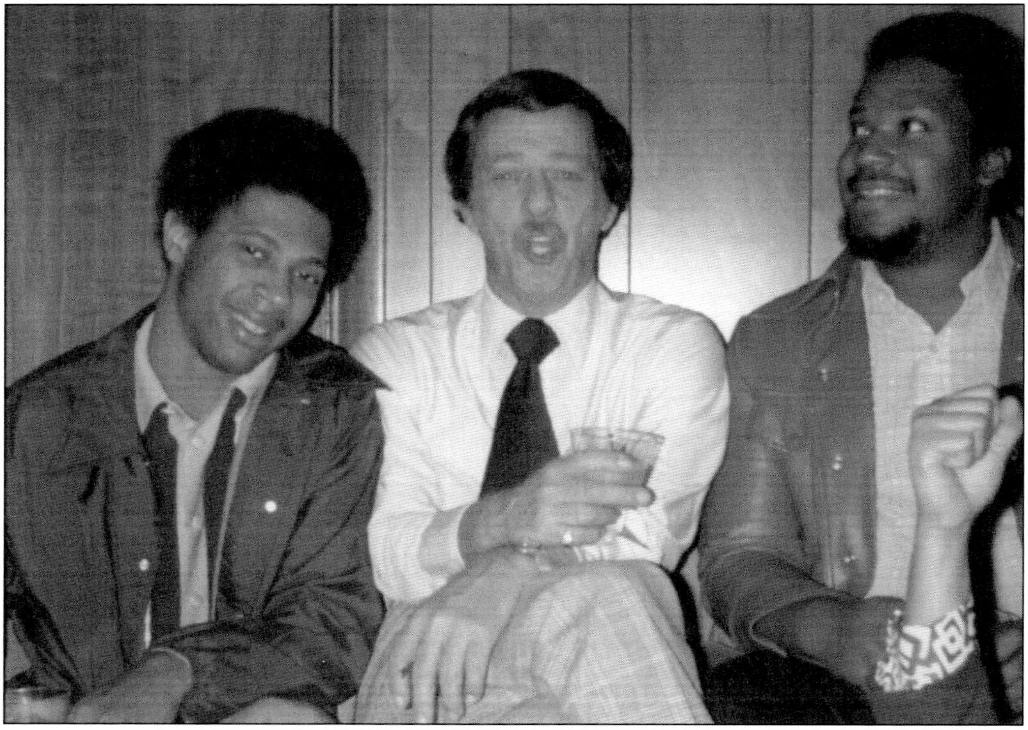

The station hosted a number of client and promotional parties at its offices on Whittle Springs Road. This photograph was taken during a lighter moment at one of them. From left to right are WNOX announcer Larry Steele, station manager Pete Dreyer, and announcer James Maurice Turk. Dreyer became manager of the station in 1967 and remained manager until 1981. Below, WNOX staff mingle with clients at a station event in downtown Knoxville. (Both courtesy of Janice McCroskey.)

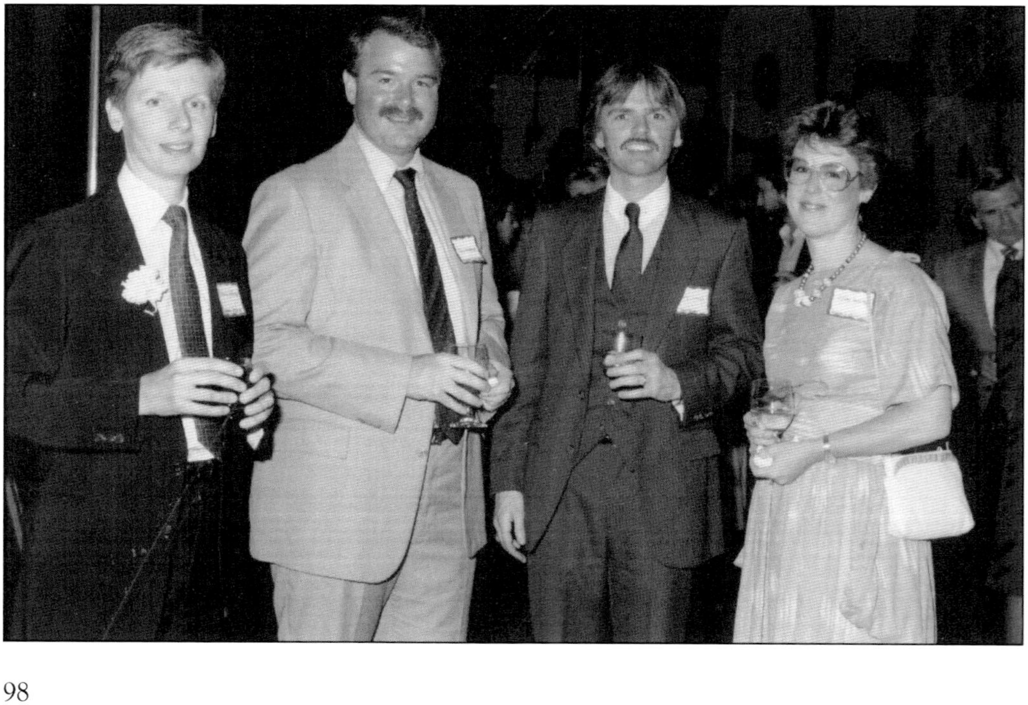

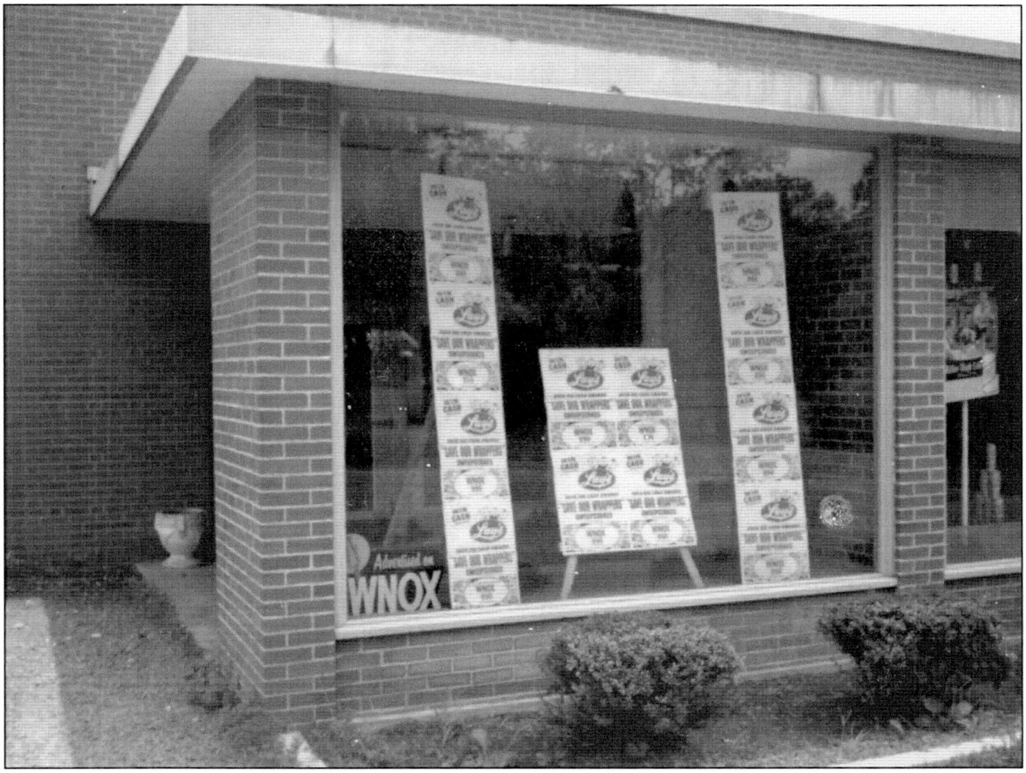

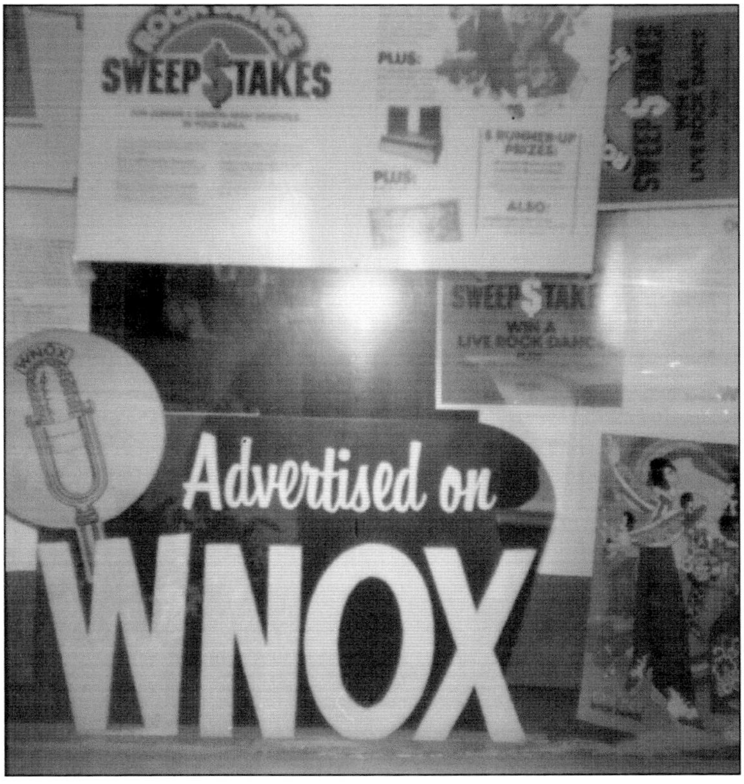

The office windows at WNOX on Whittle Springs Road were still used as they were on Gay Street. Here they were implemented as part of an advertising campaign for Lays Packing Company. While pedestrian traffic greatly diminished when the station moved from Gay Street to Whittle Springs Road, people still frequented the station enough to make the campaigns a worthwhile independence. At right is another example with the sign "advertised on WNOX." (Both courtesy of Janice McCroskey.)

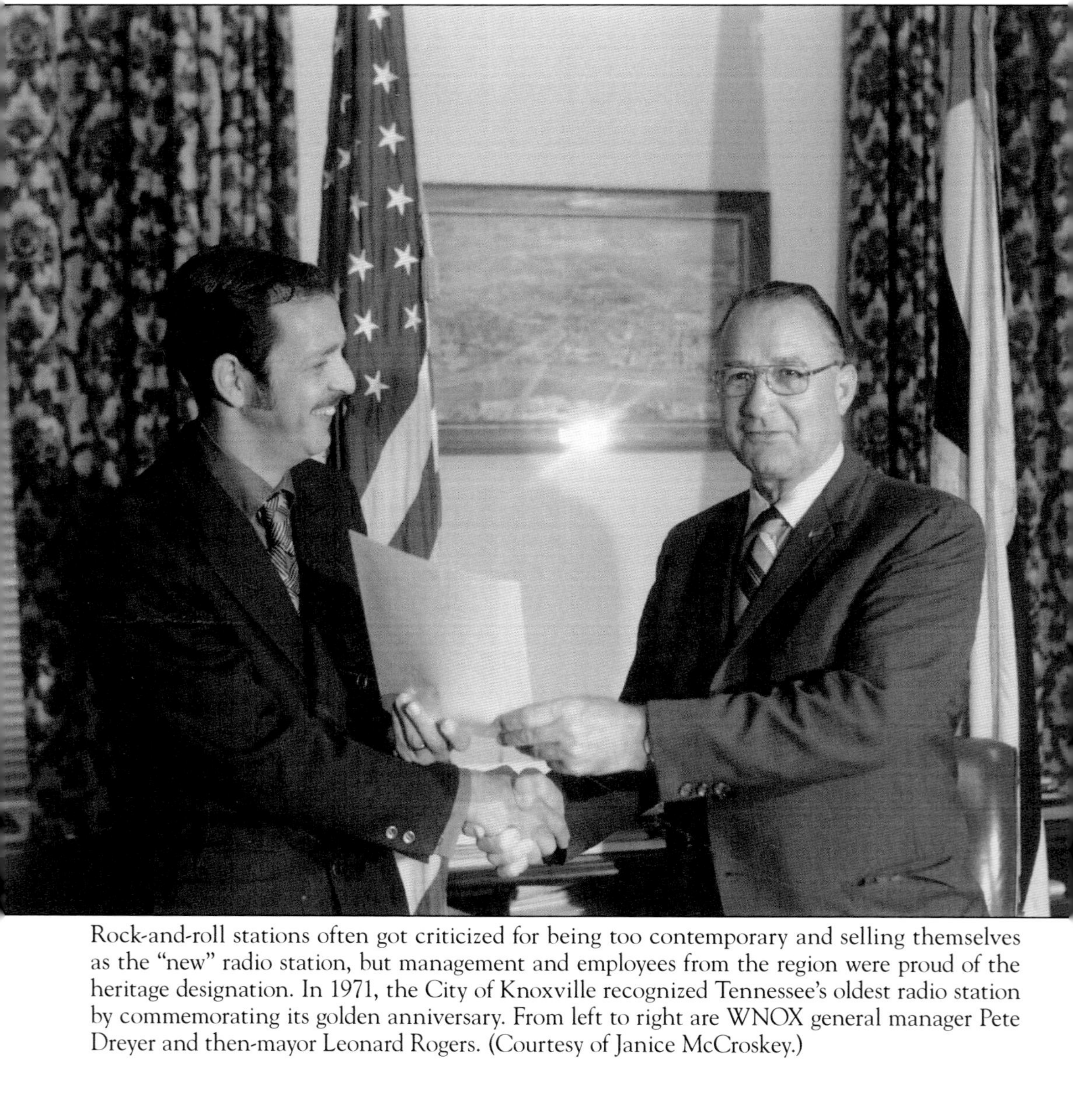

Rock-and-roll stations often got criticized for being too contemporary and selling themselves as the "new" radio station, but management and employees from the region were proud of the heritage designation. In 1971, the City of Knoxville recognized Tennessee's oldest radio station by commemorating its golden anniversary. From left to right are WNOX general manager Pete Dreyer and then-mayor Leonard Rogers. (Courtesy of Janice McCroskey.)

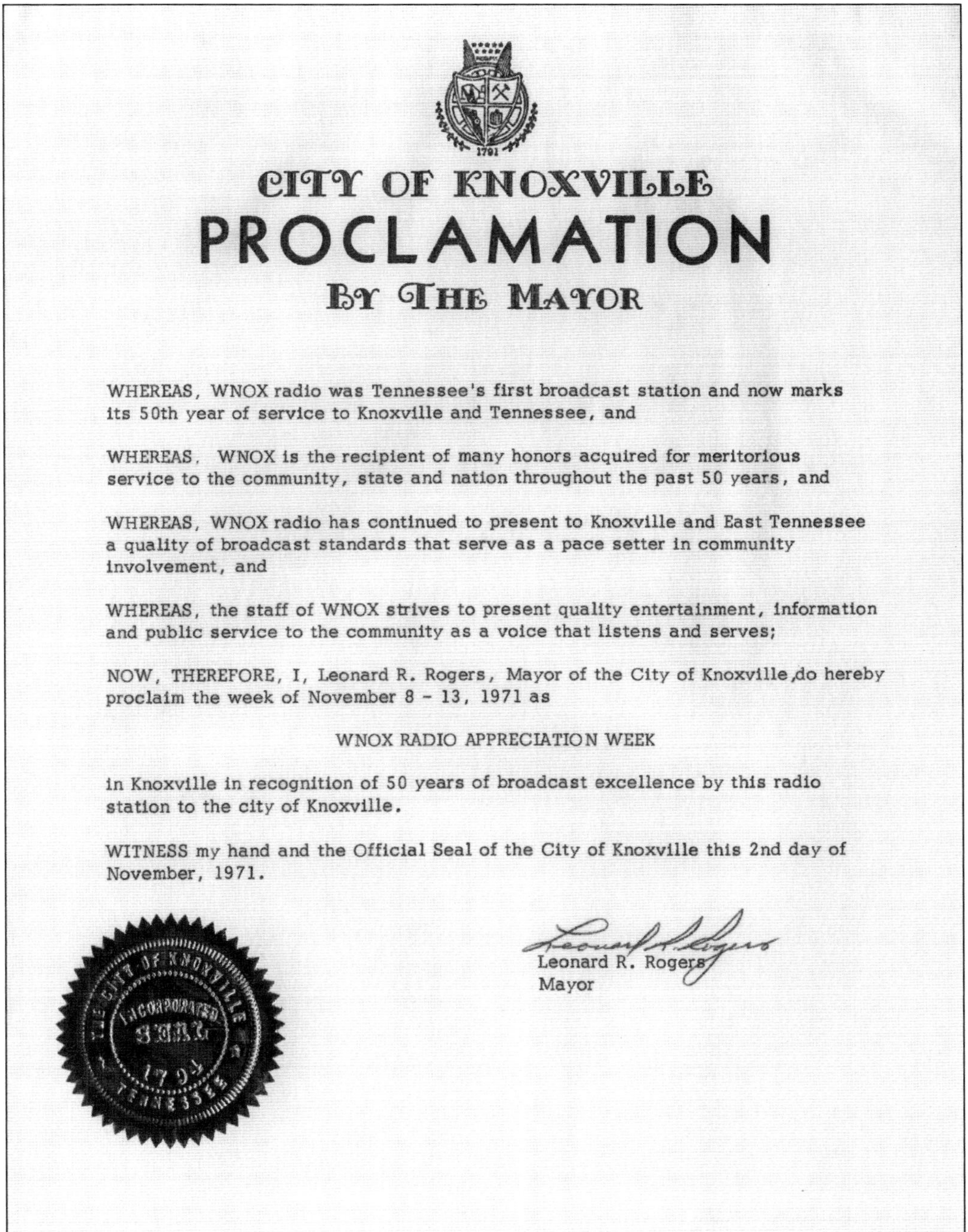

CITY OF KNOXVILLE

PROCLAMATION

BY THE MAYOR

WHEREAS, WNOX radio was Tennessee's first broadcast station and now marks its 50th year of service to Knoxville and Tennessee, and

WHEREAS, WNOX is the recipient of many honors acquired for meritorious service to the community, state and nation throughout the past 50 years, and

WHEREAS, WNOX radio has continued to present to Knoxville and East Tennessee a quality of broadcast standards that serve as a pace setter in community involvement, and

WHEREAS, the staff of WNOX strives to present quality entertainment, information and public service to the community as a voice that listens and serves;

NOW, THEREFORE, I, Leonard R. Rogers, Mayor of the City of Knoxville,do hereby proclaim the week of November 8 - 13, 1971 as

WNOX RADIO APPRECIATION WEEK

in Knoxville in recognition of 50 years of broadcast excellence by this radio station to the city of Knoxville.

WITNESS my hand and the Official Seal of the City of Knoxville this 2nd day of November, 1971.

Leonard R. Rogers
Mayor

The Knoxville Proclamation recognized the station's 50 years of broadcasting by proclaiming November 8–13, 1971, as WNOX Radio Appreciation Week.

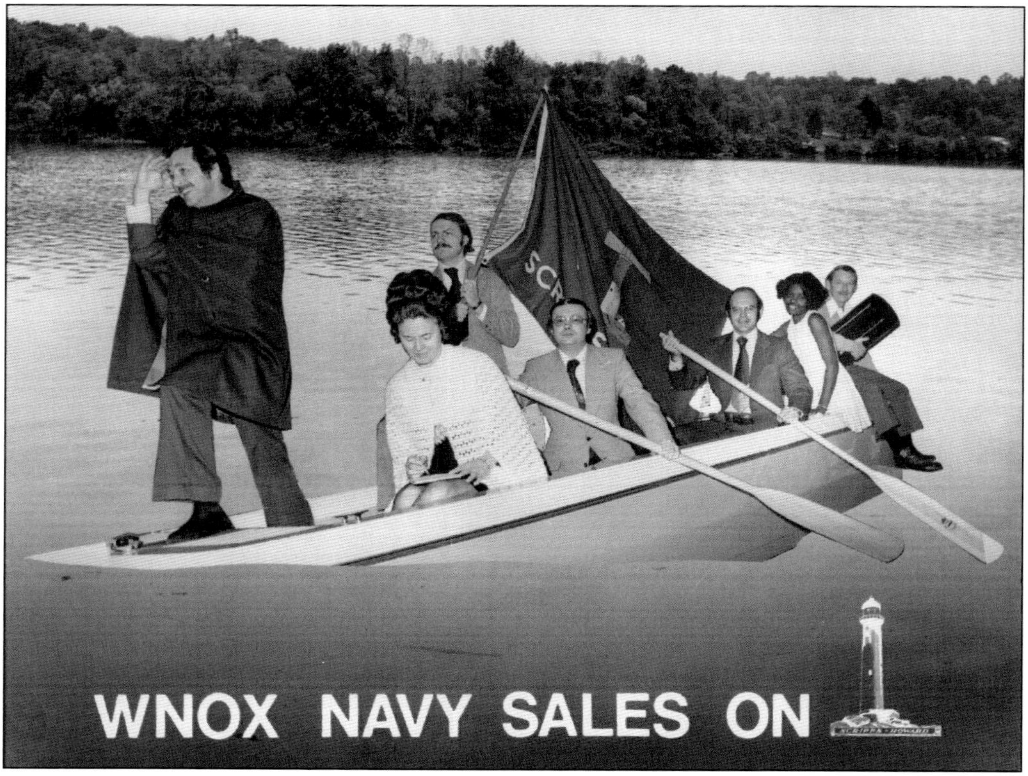

This promotional photograph was taken of the WNOX sales staff. From left to right are Pete Dreyer, Janice McCroskey, Eddie Roy (holding the Scripps Howard flag), Bob Byerley, Bill Hayes, unidentified, and Clay Wright. (Courtesy of Janice McCroskey.)

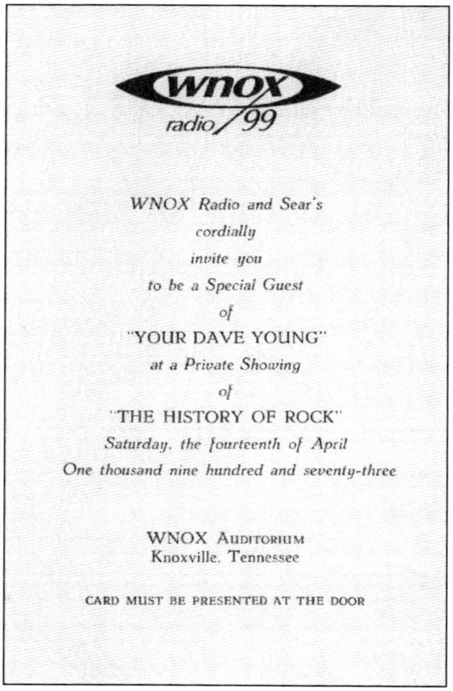

wnox radio/99

WNOX Radio and Sear's
cordially
invite you
to be a Special Guest
of
"YOUR DAVE YOUNG"
at a Private Showing
of
"THE HISTORY OF ROCK"
Saturday, the fourteenth of April
One thousand nine hundred and seventy-three

WNOX Auditorium
Knoxville, Tennessee

CARD MUST BE PRESENTED AT THE DOOR

In 1973, Sears Department Stores and WNOX did a joint promotion on "The History of Rock" with WNOX announcer Dave Young. The promotion was held in the station's auditorium as an educational outreach for Knoxville and Knox County Public Schools.

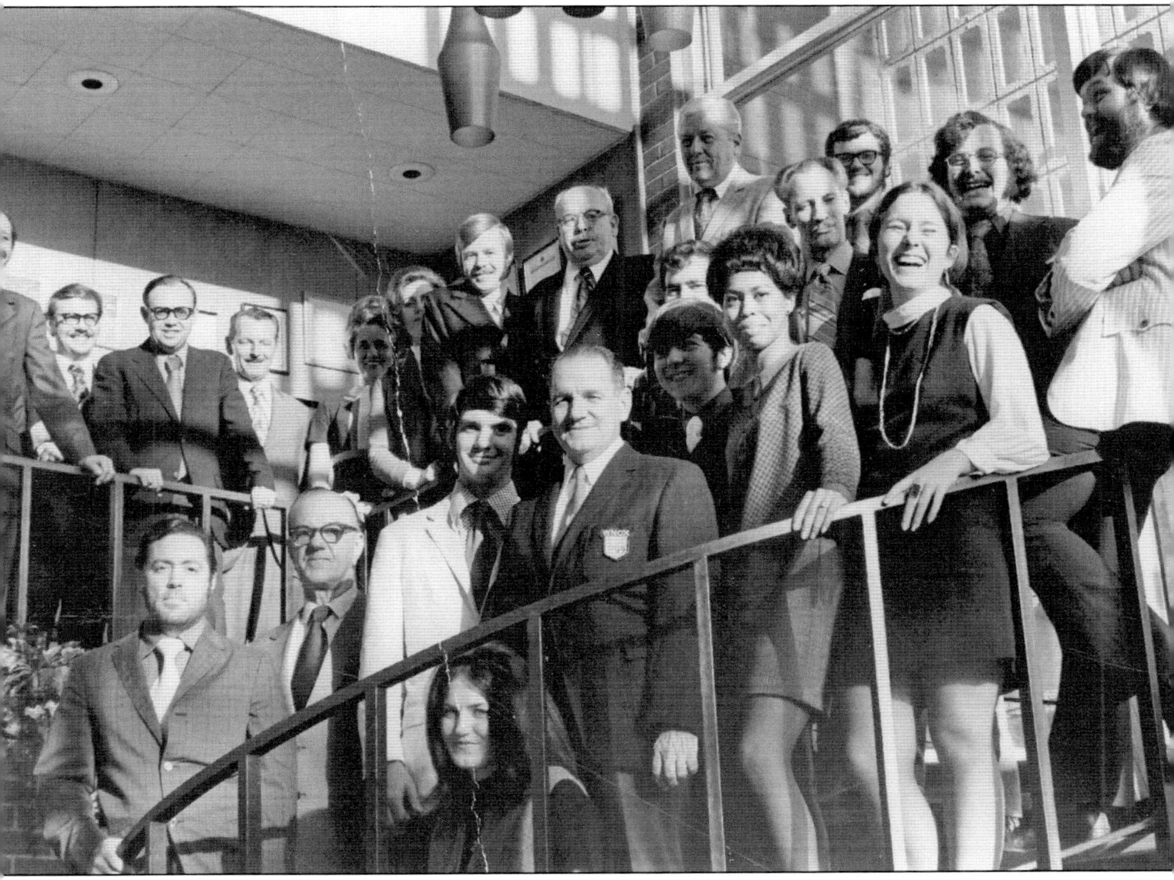

The WNOX staff poses for a photographer in this c. 1971 image on the circular stairway that sat in the lobby of the station's Whittle Springs office. Those identified from left to right going up and then to the left are John Maples, Bob Engert, Terry Blankenship (in front), Lee Benedetti, Dennis Shehan, Pat Thompson (in shade), unidentified, Carolyn Vreelan, "Skinny" Kenny Mynatt, Fleetwood Gruber, Ron Ashburn, Raymond Stecker Wilk III, Fred Shepherd, Hubert Carter, unidentified, Your Dave Young, Pat Lemons, Jan McCroskey, Clay Wright, Bob Byrley, Eddie Roy, and station general manager Pete Dreyer. (Courtesy of Janice McCroskey.)

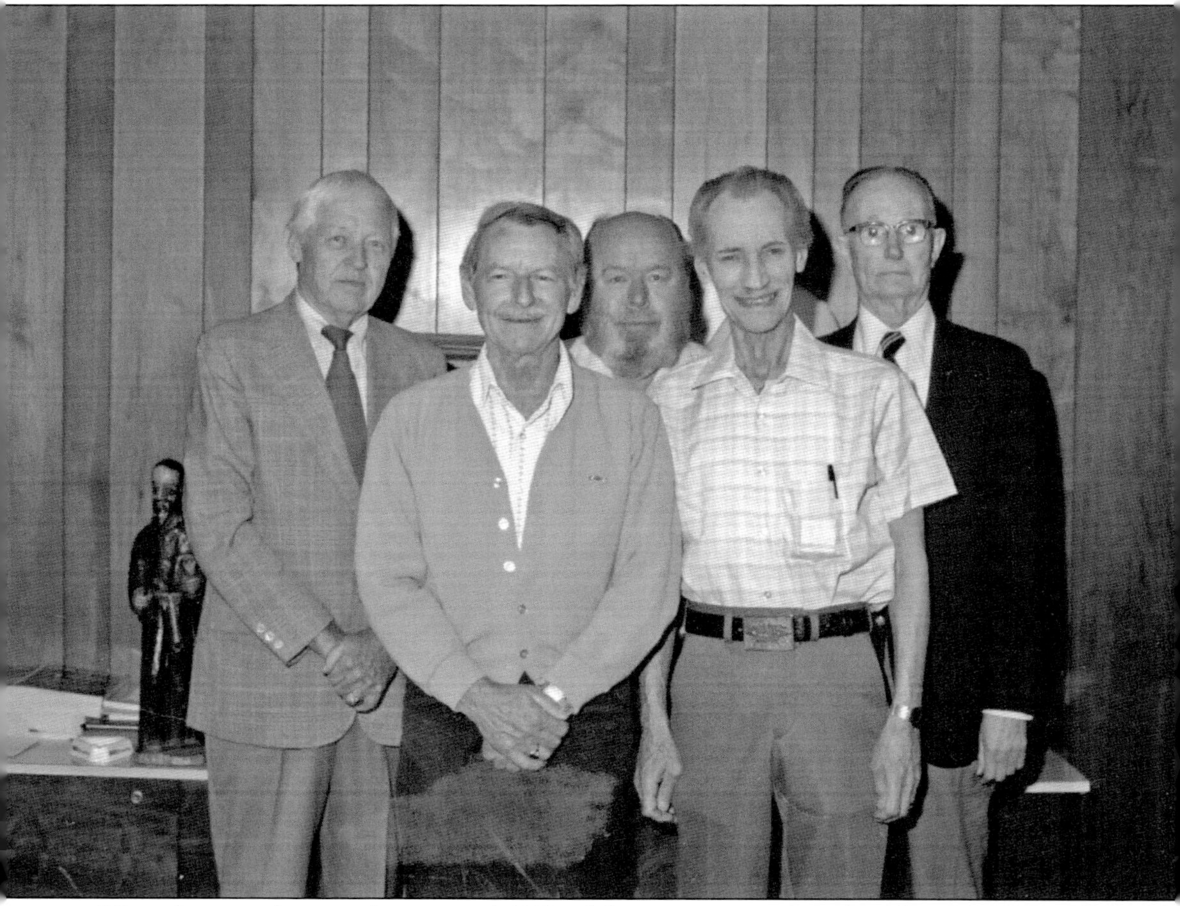

This photograph was taken around 1971 of the remaining original staff at WNOX as they celebrated 50 years of broadcasting in Tennessee. From left to right are Fred Shepherd, Clay Wright, Hubert Carter, Ron Ashburn, and O. L. Smith. In 1972, the longest serving employee, Fred Shepherd, retired from WNOX/Scripps Howard and ventured back on a part-time basis for a few years afterward. (Courtesy of Janice McCroskey.)

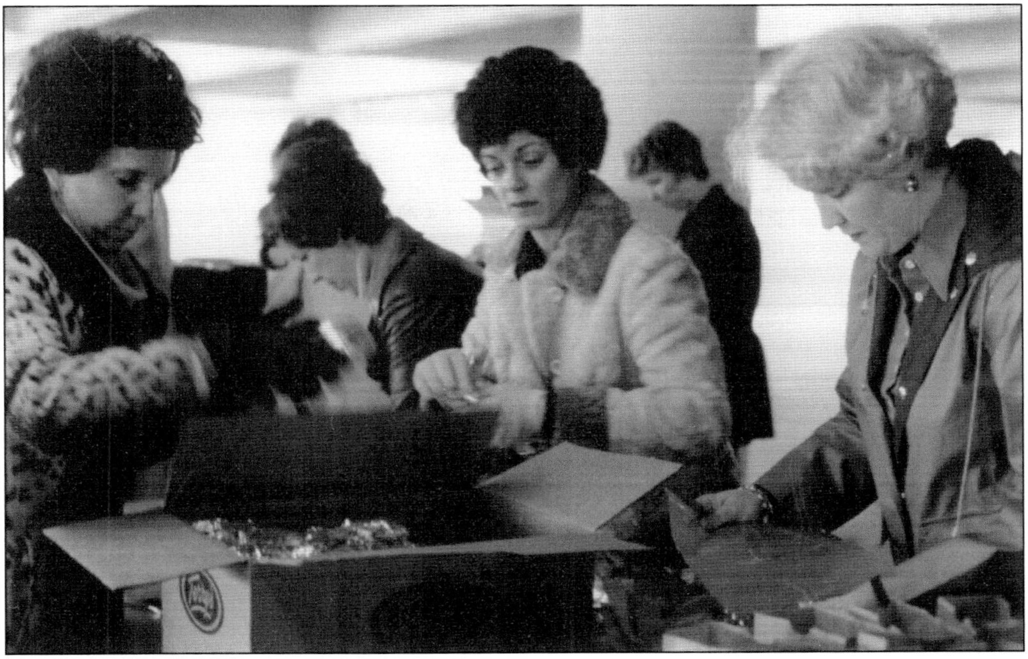

WNOX also did a variety of special needs projects through the years to help the community's less fortunate. Under Scripps Howard, the station often assisted the *Knoxville News Sentinel* in its annual Empty Stocking Fund in addition to its own programs. During the holidays, they participated in a number of projects. Here ladies work to put together baskets for one of the charities. Below is an aluminum can drive done in cooperation with the Cherokee Distributing Company. (Both courtesy of Janice McCroskey.)

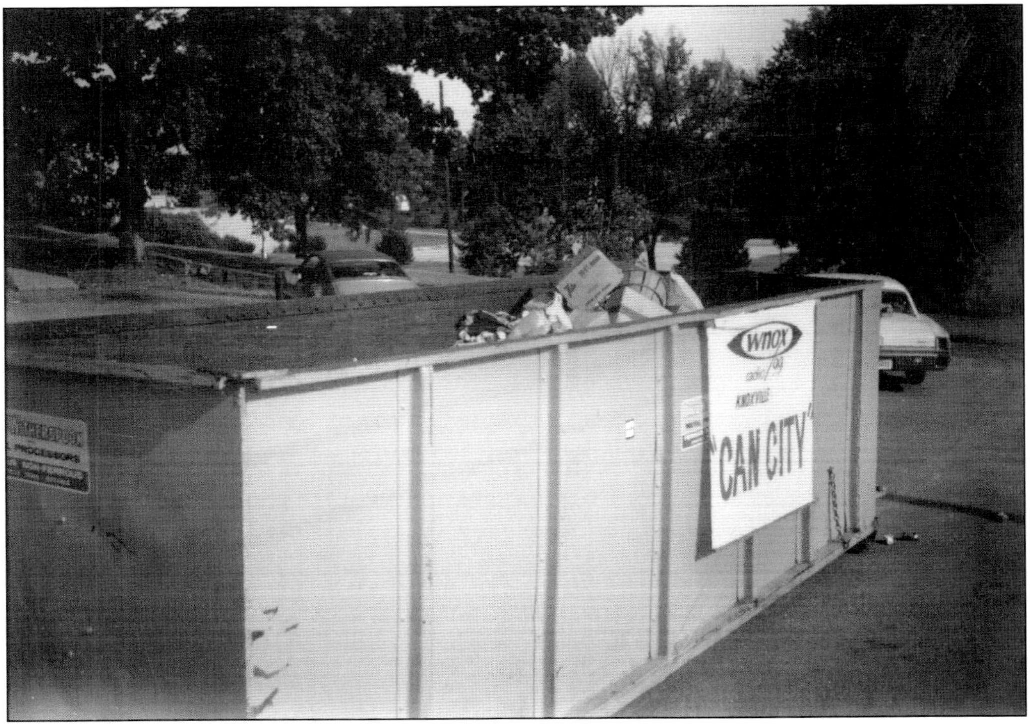

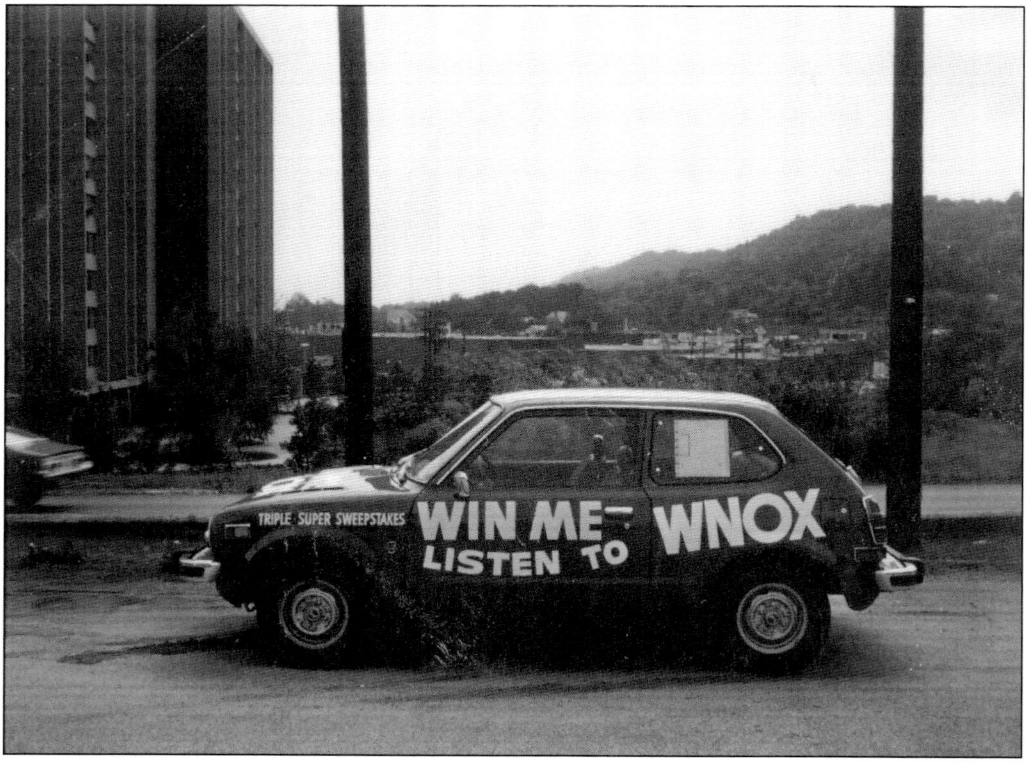

During the gas wars and oil crisis of the 1970s, WNOX ran a very successful promotion giving away a fuel-efficient Honda automobile. With gas at the then-outrageous price of $1.60 per gallon, it was one of the most popular station promotions ever. (Courtesy of Janice McCroskey.)

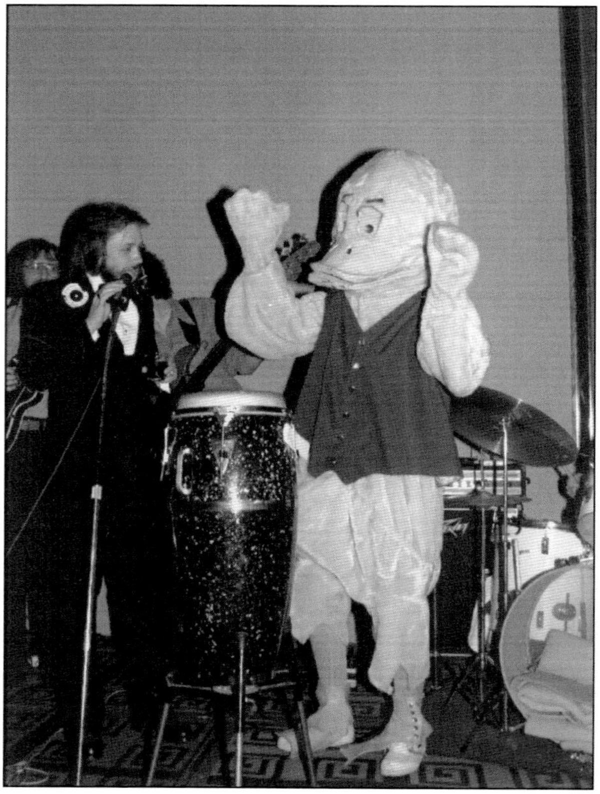

The business mascot fad of the 1970s with the San Diego chicken also led WNOX to come up with the duck as its official station mascot. Here Eddie Beacon ("the Swinging Deacon") of WNOX 1960s fame introduces the mascot to the audience. (Courtesy of Janice McCroskey.)

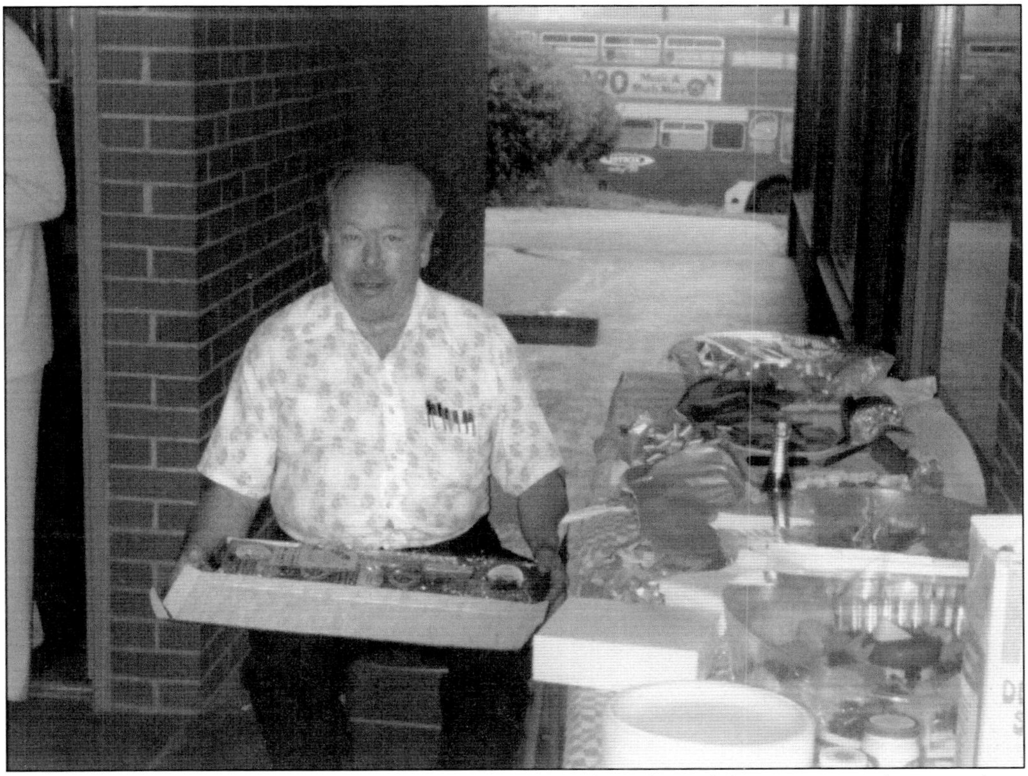

Hubert Carter worked in traffic at WNOX from the earliest days of the station, working as a bass player at the *Mid-Day Merry-Go-Round* and *Tennessee Barn Dance*. Above, he is given a retirement party by the station. Behind Carter is a double-decker bus WNOX used as a promotion vehicle for station events. Below, he is pictured with WNOX general manager Pete Dreyer. (Both courtesy of Janice McCroskey.)

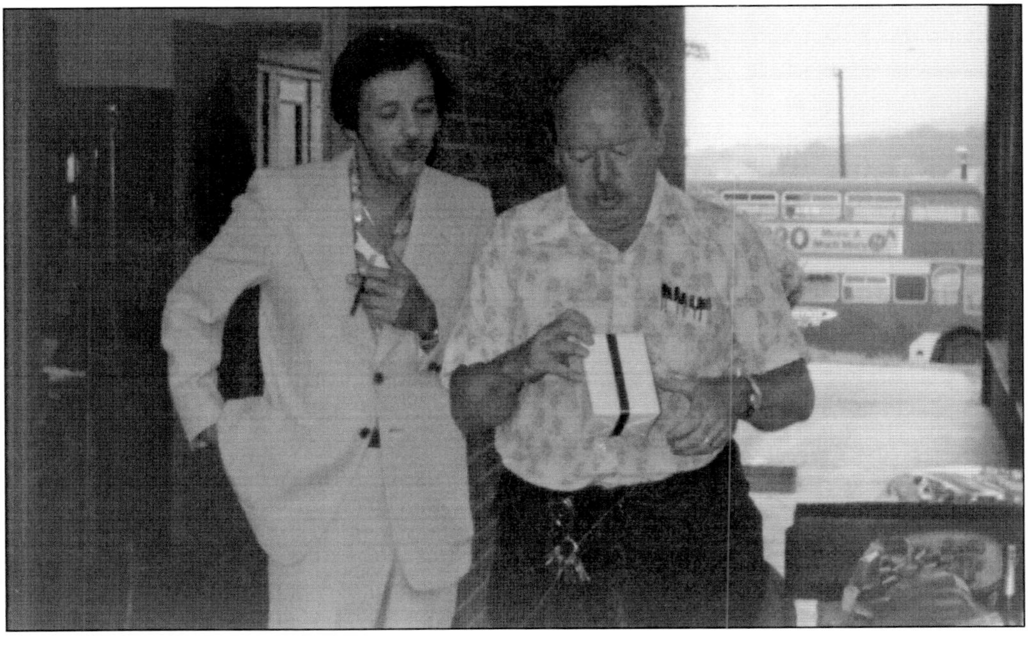

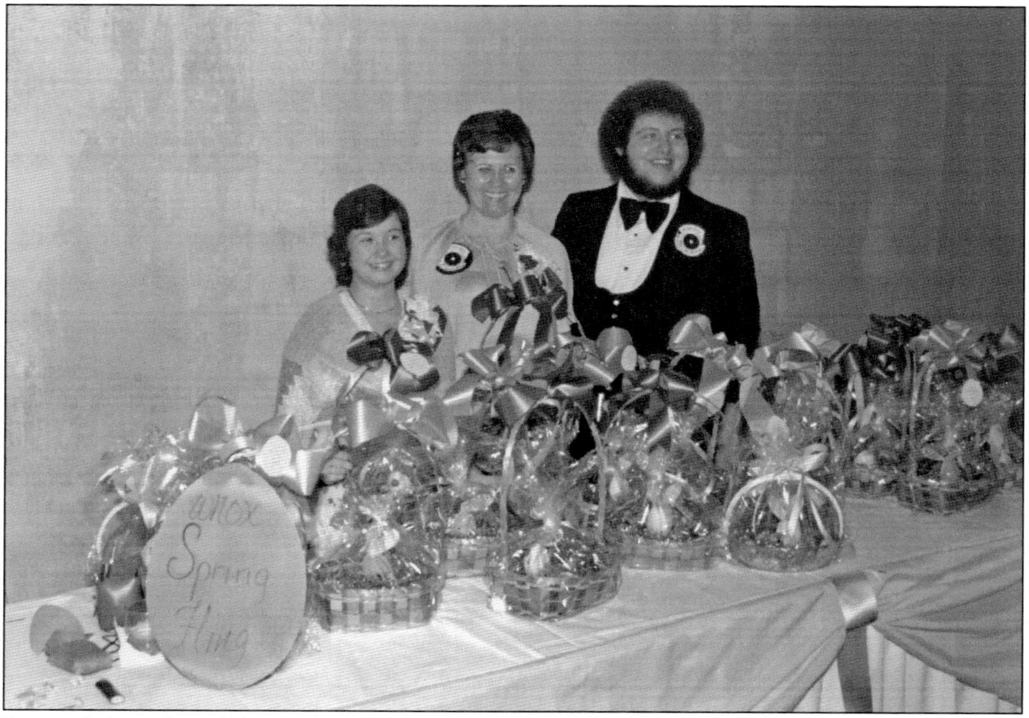

Pictured above from left to right are Jan Oliver, Jan McCroskey, and Phil Williams during WNOX's annual Spring Fling. Jan Oliver now works in traffic at WIVK, and Phil Williams would become one of Knoxville's most popular morning drive announcers at WIMZ before landing back at WNOX, where he does the afternoon program *The Phil Williams Show*. The annual Spring Fling was a big event for the station. Pictured below from left to right are Trudy Dreyer, Ron Ashburn, and Mrs. Ashburn (Jane). (Both courtesy of Janice McCroskey.)

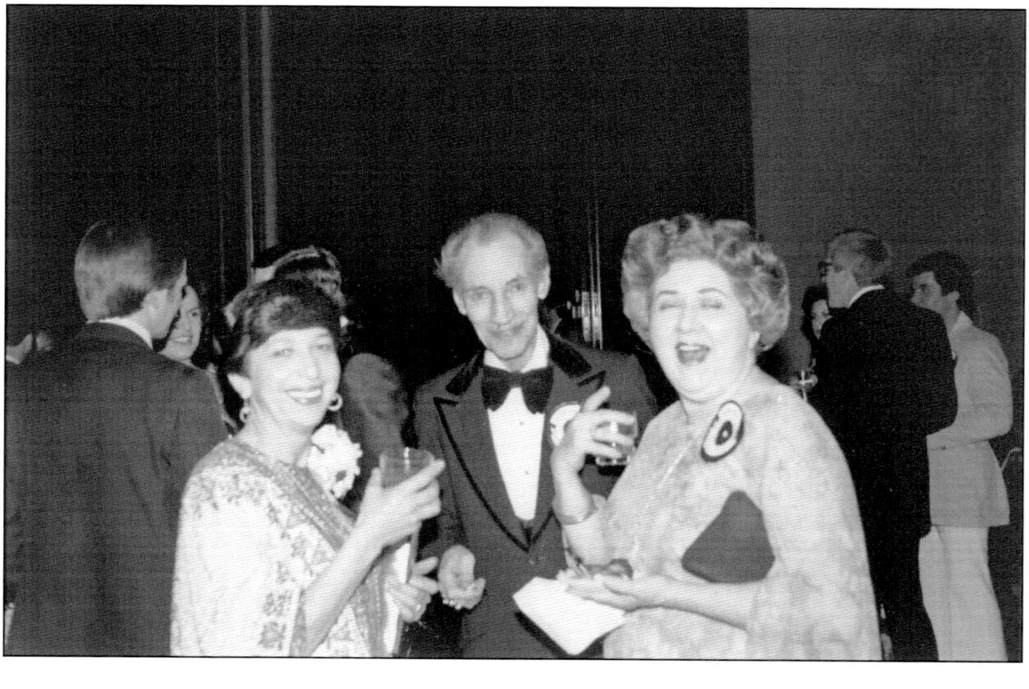

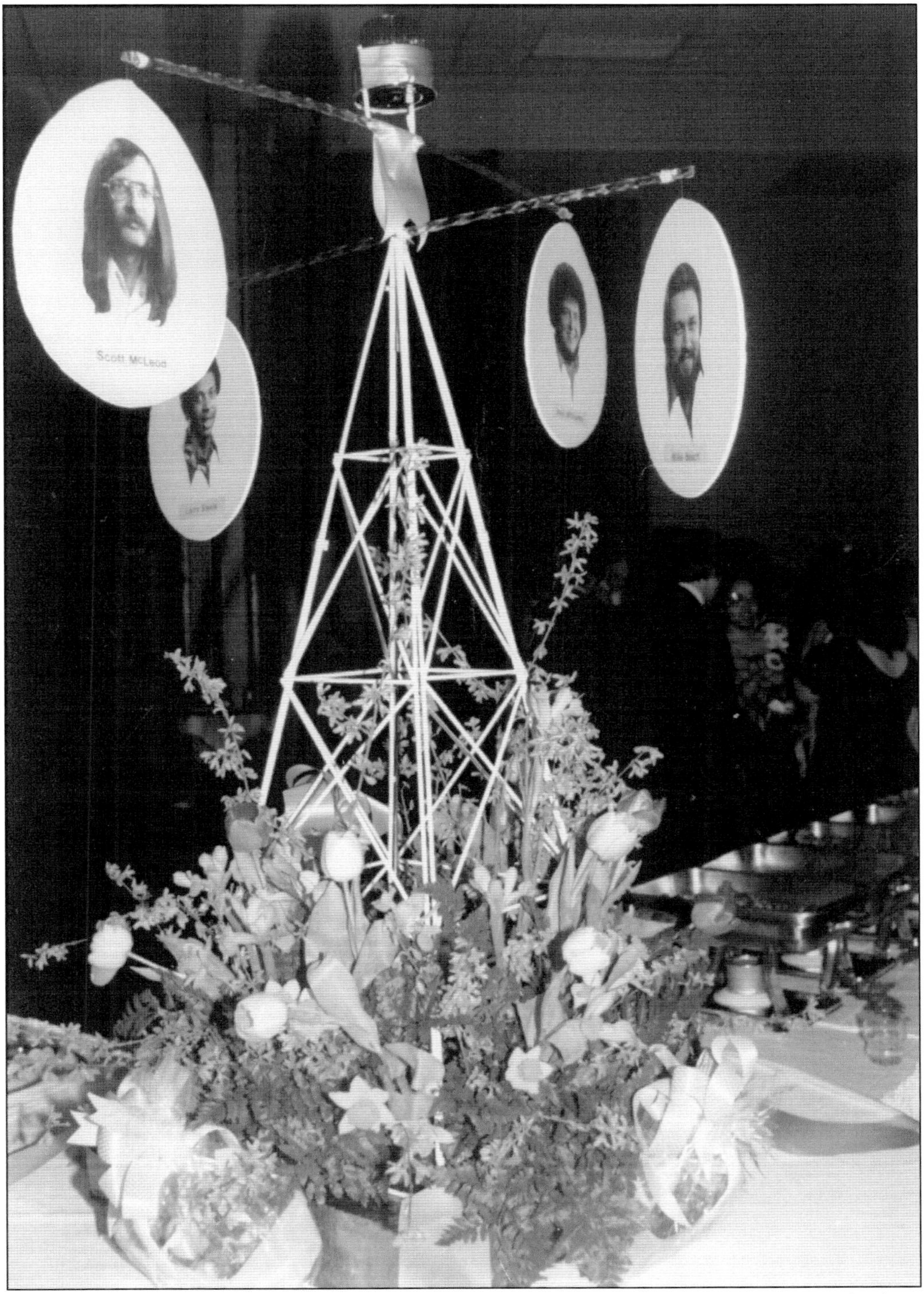

The Spring Fling was also a good way to promote the station to guests, advertisers, and others in attendance. WNOX's creative centerpiece for the event was a transmitter tower with cameo photographs of the station's announcing crew. (Courtesy of Janice McCroskey.)

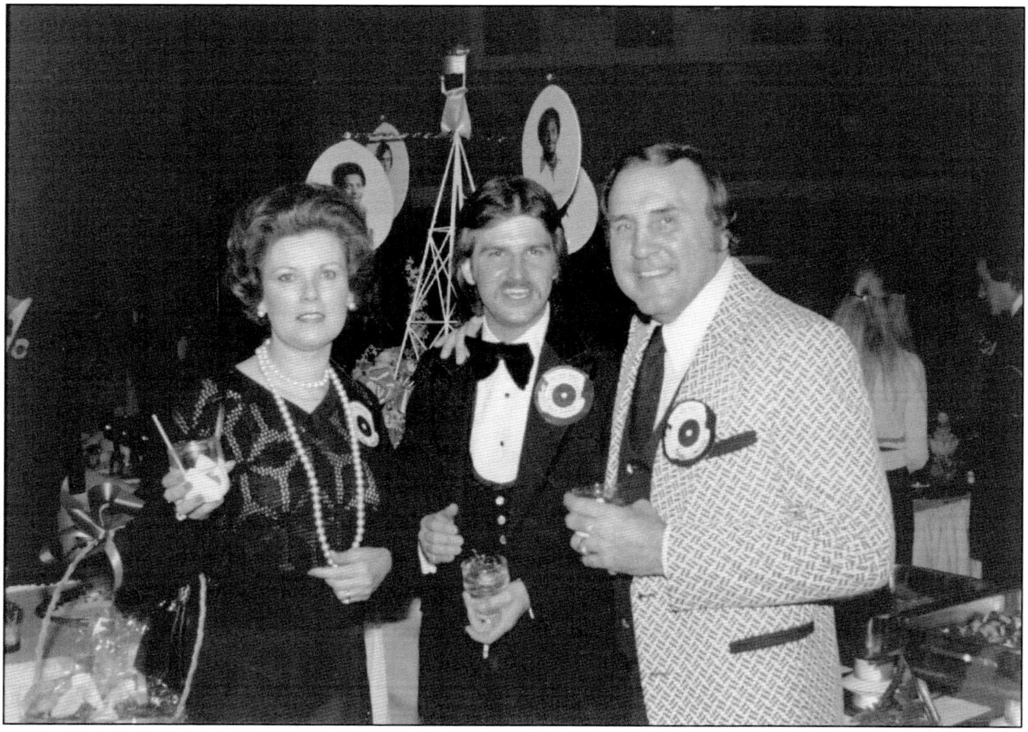

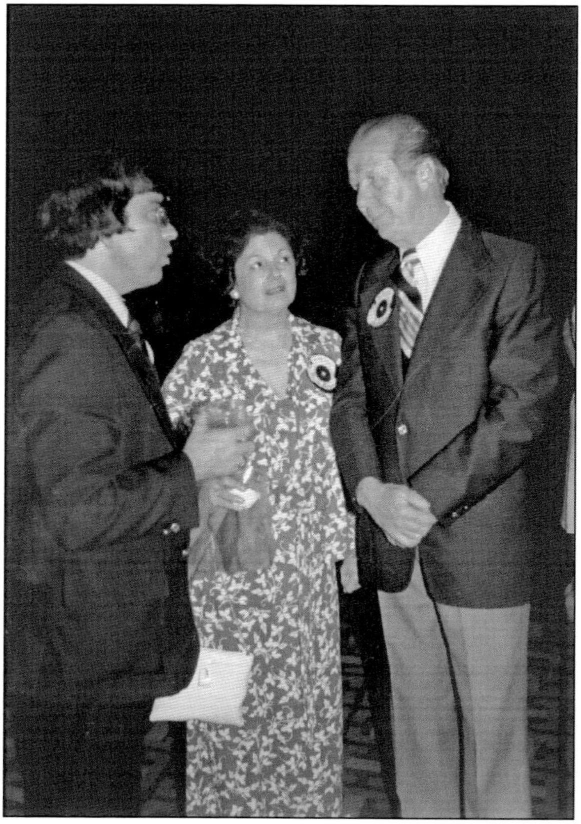

At station events, the entire staff always turned out. Unless they were scheduled to work, station personnel were required to attend. Here the staff mixes with the crowd. Pictured with this unidentified woman are morning announcer Steve Land (center) and Andy Smalls, who is better known in Knoxville as Marshal Andy. (Courtesy of Janice McCroskey.)

Knoxville city councilman M. T. "Tee" Bellah (far right) speaks to a WNOX employee. Bellah had a history with WNOX. He was a gifted pianist, and he and his brother Jim had a weekly radio program where the brothers played dual pianos. Bellah passed away in 2008 at the age of 94. (Courtesy of Janice McCroskey.)

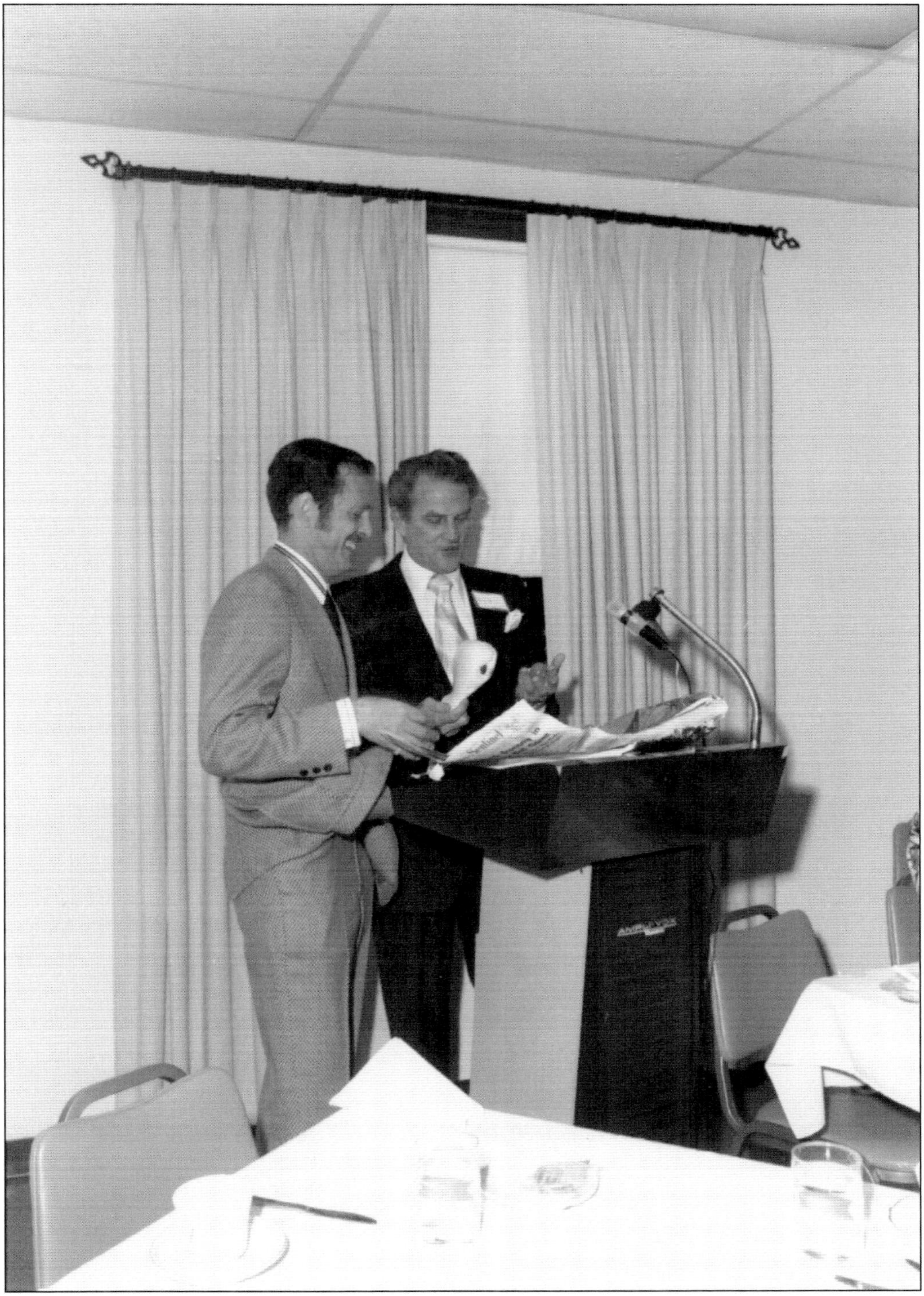

In the 1970s, roasts were popular entertainment where people gathered and joked about the guest of honor. WNOX held one for Pete Dreyer and gave some of the leading movers and shakers in Knoxville a chance to take their best verbal shot at the station's general manager. Here the *Knoxville News Sentinel*'s Bill Burke (right) presents a gag gift to Dreyer. (Courtesy of Janice McCroskey.)

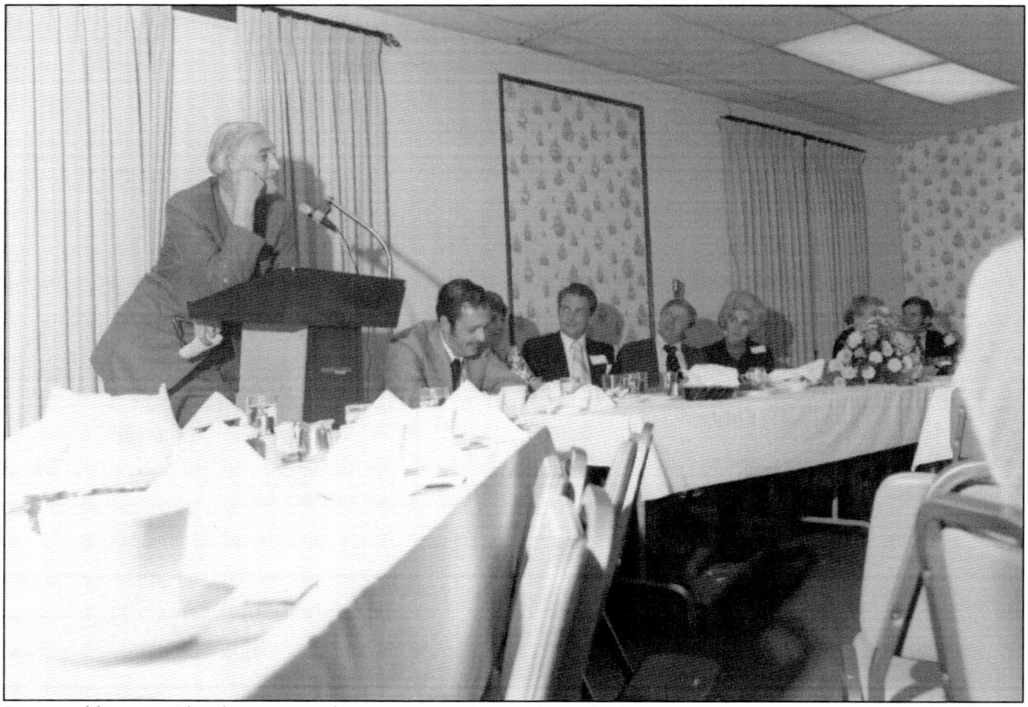

Pictured here is Charlie Davis of Davis Newman Payne advertising roasting Dreyer. The advertising agency had a long history with WNOX going back to the early days of the station. They packaged Archie Campbell from the *Mid-Day Merry-Go-Round* and sold his stage bits to Chevrolet Dealers across the nation as a package to use in radio commercials, which launched Campbell to fame in America. Below is an image of Liz Tombras with her turn at the microphone. (Both courtesy of Janice McCroskey.)

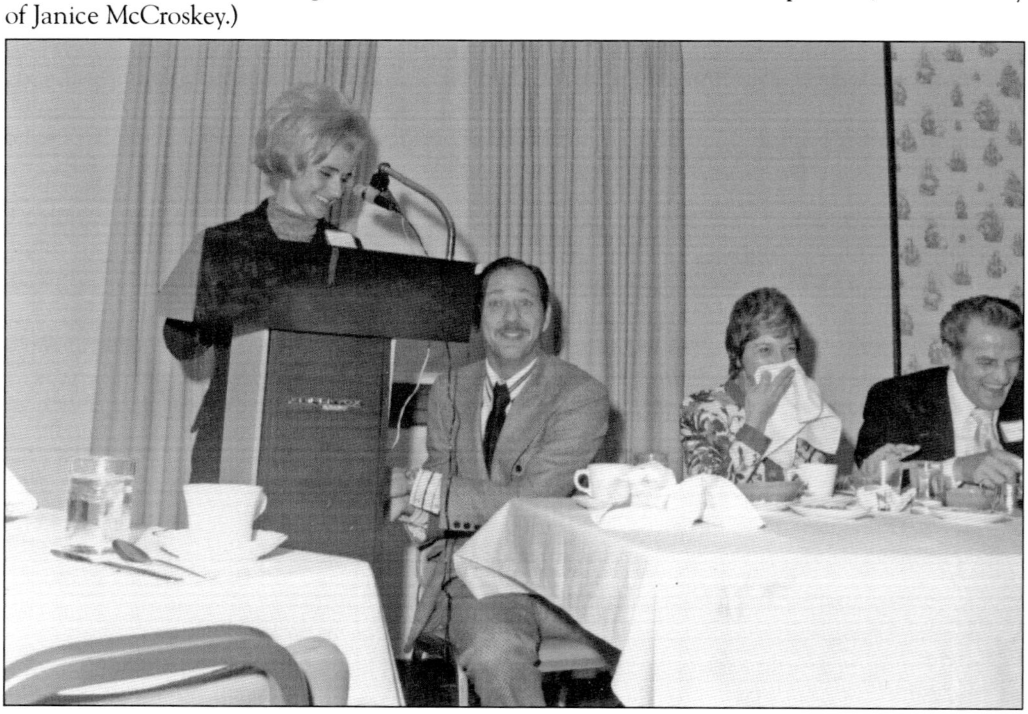

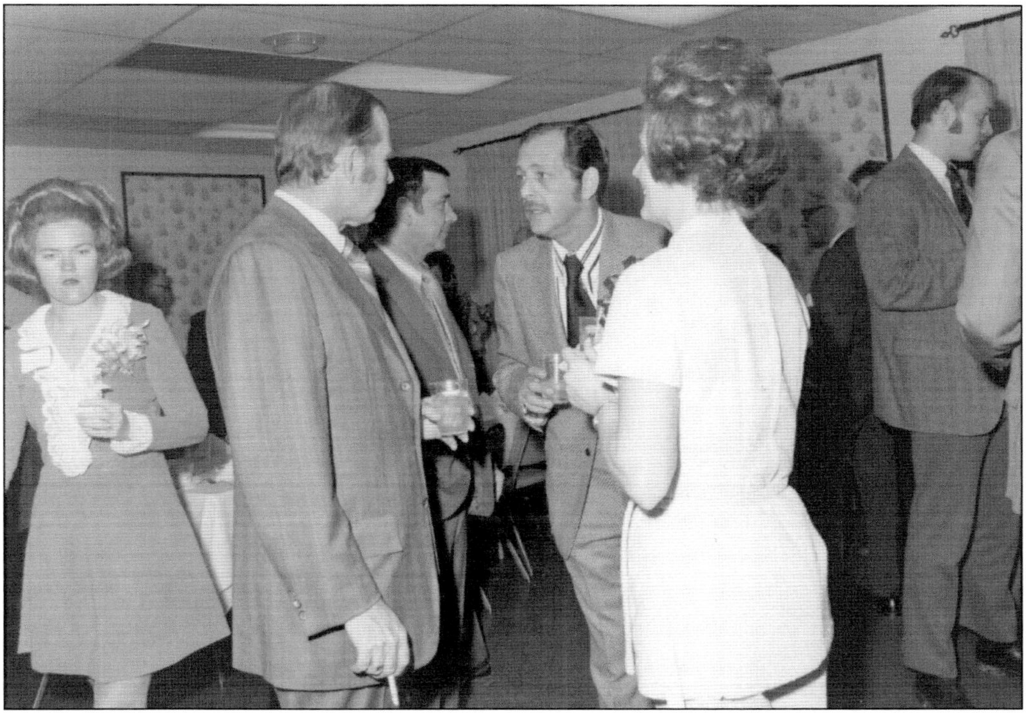

WNOX general manager Pete Dreyer mingles with guests at the roast. From left to right are Betty Byerley, Bill Eckstein, unidentified, Pete Dreyer, and Jan McCroskey. Below, he poses with the roasters at the event. From left to right are Charlie Davis, Liz Tombras, Ted Brosseau, unidentified, Pete Dreyer, Trudy Dreyer, and a man identified only as "the Coca-Cola representative." (Both courtesy of Janice McCroskey.)

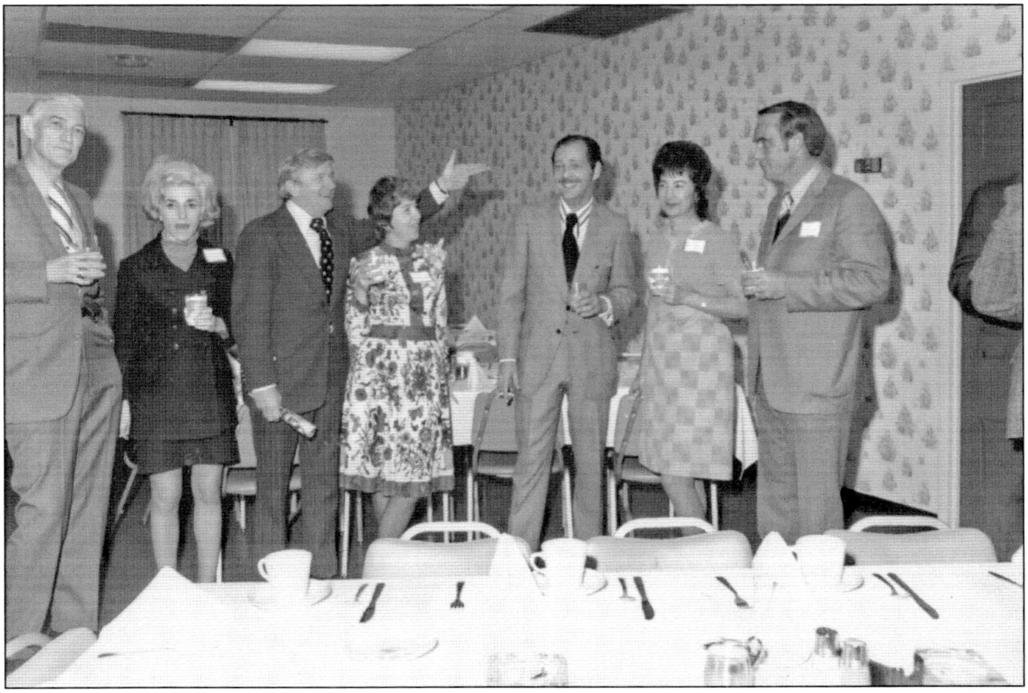

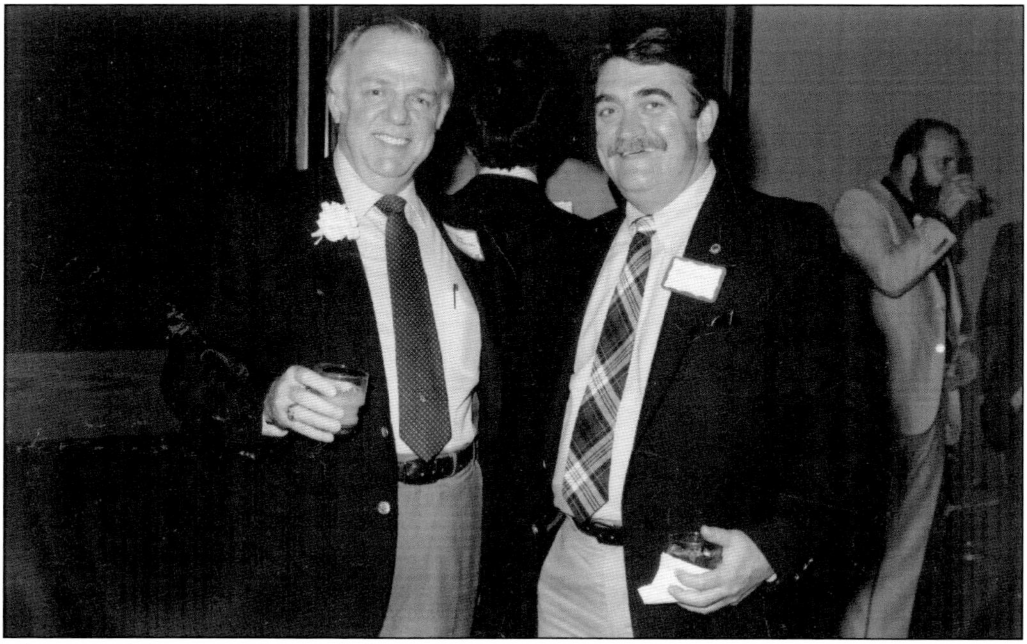

This is Paul Oscar Anderson (left), who took over WNOX news when longtime anchor and news director Ron Ashburn retired from the station. Filling the shoes of one of East Tennessee's premier newsmen was a tough job, but Anderson's big booming voice quickly became the new trademark sound of WNOX news. The billboard below was a promotion run by the station to push his newscasts. (Both courtesy of Bill Hayes.)

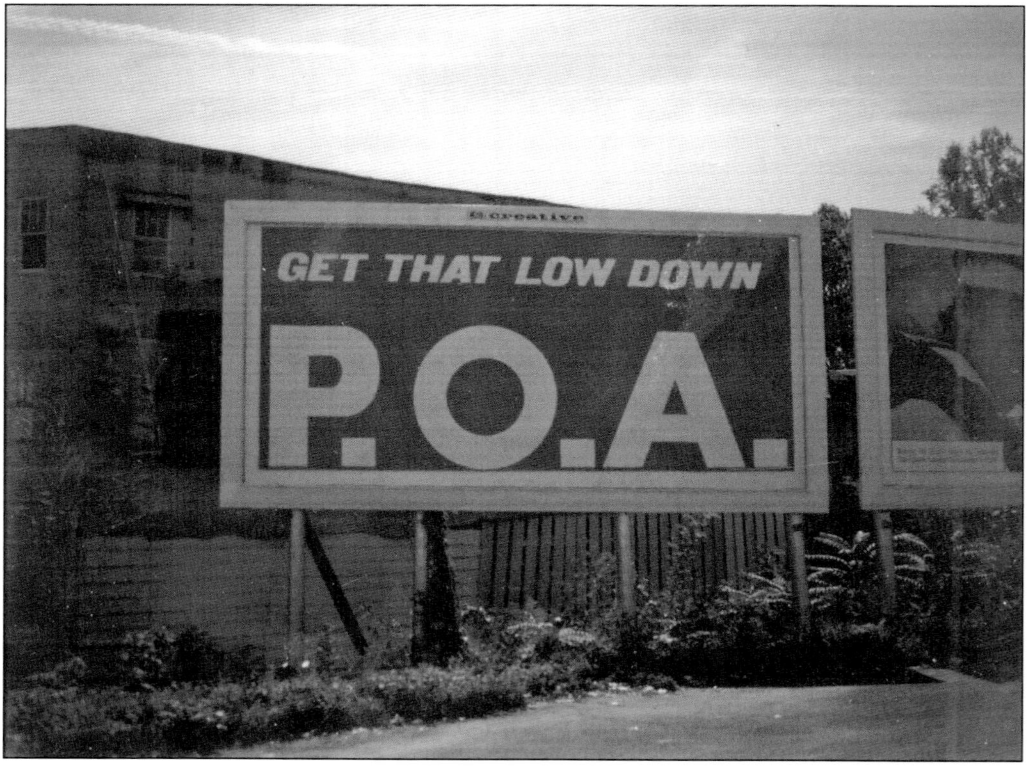

Six

Back to Country and Beyond

In 1982, Scripps Howard divested itself of WNOX after more than 40 years of continuous ownership and sold it to Mack Sanders, who quickly switched the station back to a country music format. He also bought other stations in East Tennessee and tried to position WNOX as a powerhouse against the reigning station, WIVK. The auditorium was put back into the mix, and a huge promotional campaign was kicked off to sell the concept to listeners. With Scripps Howard out of the picture, many broadcasters and former employees would say the station was never the same again. Scripps Howard Broadcasting left a footprint in Knoxville and East Tennessee that is legendary.

In 1985, the station was sold again to REBS Communications, who sold it almost immediately to ELF Communications, which was owned by John Lack, one of the founders of the Music Television network. ELF Communications took off with WNOX and, like Sanders, tried to bring Tennessee's oldest radio station back to its roots as the founding station of country music. Lack rededicated the WNOX Auditorium with great fanfare and an audience filled with former *Mid-Day Merry-Go-Round* performers and Nashville's elite country music recording artists. They unveiled the new Lowell Blanchard Auditorium and hosted an inaugural concert for it. By then, WIVK was firmly entrenched among listeners and pulling in rating numbers that did not give WNOX a real chance for dominating like it had throughout its history. It maintained its news and public service commitments and had some of the same voices, which kept its ratings respectable. The revived *Tennessee Barn Dance* was a huge plus for the station, and media promotion was never difficult to find for the legendary institution.

Despite the inclusion of old names and faces from WNOX's past, the station could not recover the mass appeal it had traditionally enjoyed with East Tennessee listeners. It stayed alive long enough to give a new set of performers the start they needed and refurbish a couple of old acts in need of revival. In 1987, however, it went into receivership, and WNOX on Whittle Springs Road was forced to close its doors. It was a complete surprise to broadcasters and listeners alike that the 66-year-old station that had opened the door to the world for generations of East Tennesseans was no more. It fell prey to the business of radio, and it can be said that the very industry WNOX had helped pioneer was now turning on it and cannibalizing the heritage station. Its call letters were a trophy, and people tried to claim its heritage without the same work and effort it took to build and maintain a number one country and subsequent number one rock-and-roll station.

A year later, WNOX 990 AM was purchased by Dick Broadcasting, who owned WIVK. They took the signal on the dial and released the call letters. For broadcasters and many East Tennesseans, it was unbelievable that one of the most known call letters in the country were released from 990 AM, which became a simulcast of WIVK.

The call letters were picked up by radio station owner Chuck Ketron in Sevierville and assigned to 99.3 FM in Jefferson City. Art Dees was general manager, Al Palmieri was sales manager, and this author, Ed Hooper, was hired from Knoxville to serve as news director and station news anchor. The charm of the station was gone. It was Hooper's second time with WNOX, and it felt strange to work and speak the call letters outside the city of Knoxville.

The letters remained there until they were reacquired by Dick Broadcasting in 1997. The WIVK simulcast was severed, and the signal was reunited with WNOX. It remained that way for a few years as 990 AM joined the ranks of other AM stations in becoming a news/talk format. It then became News Talk 99 when the format was simulcast on 99.1 FM.

In 2005, the WNOX call letters once again moved to 100.3 FM, which had been a former Top 40 station. The 990 AM signal was made into an all-sports station. The call letters are once again a dominant force in Knoxville broadcasting with a completely different format from its history. The News Talk station is regarded as one of the best radio news sources in East Tennessee and has been the recipient of many prestigious news awards over the years for its coverage of local events.

As with all broadcast legends, there is irony in the story. Phil Williams, who is a former announcer of the station in its rock-and-roll era, now hosts WNOX's popular afternoon talk show. The 100.3 FM radio station where the WNOX call letters now rest was once known as WOKI. It is the Top 40 radio station founded and owned by former WNOX music director John Pirkle, who was fired from his position at the station in 1971 for starting the competing enterprise.

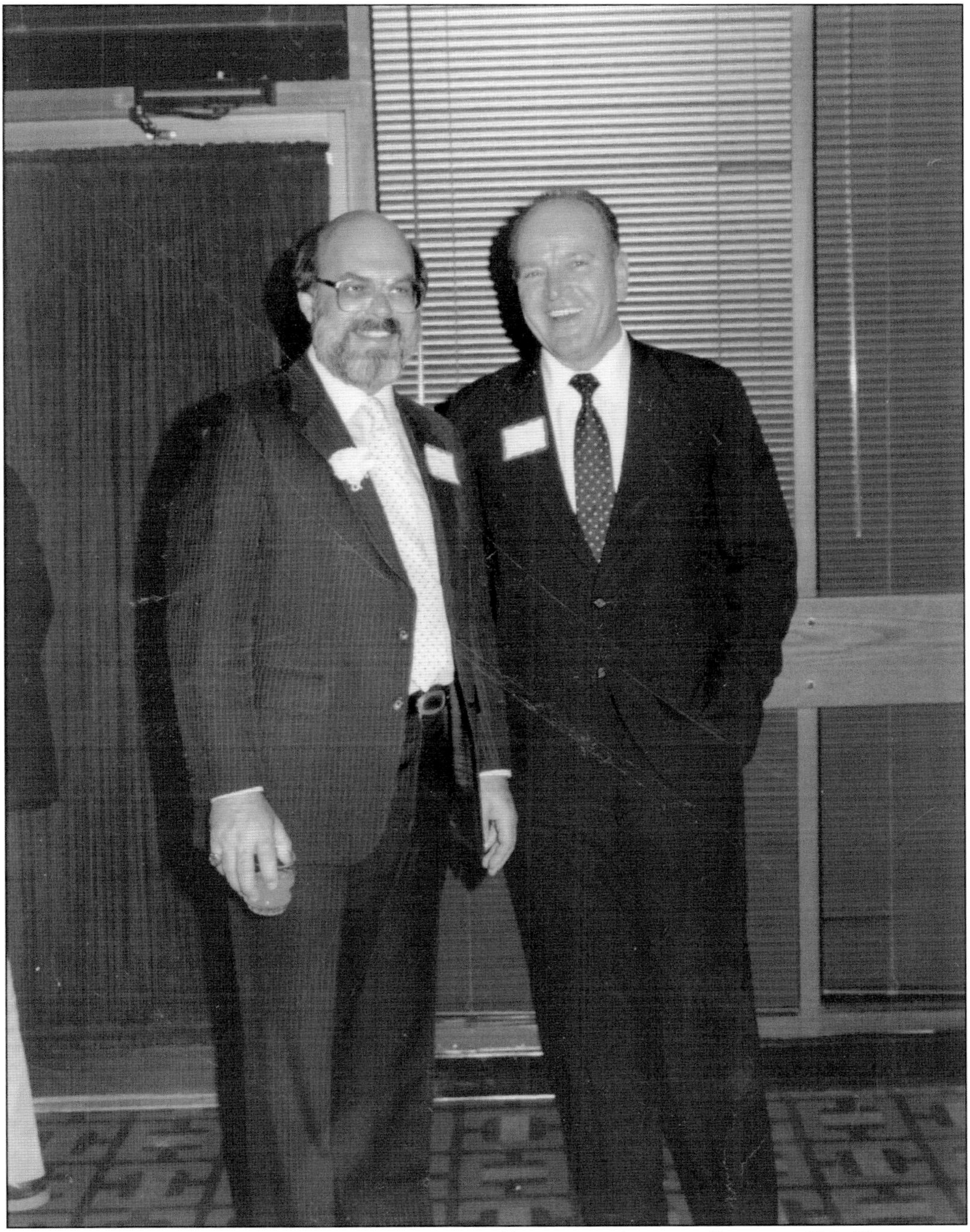

From left to right are WNOX general manager Bill Hayes and Knoxville mayor Kyle Testerman. The station's jump back to a country music format was welcomed by many. They were pleased to know the station would still maintain its community commitment and be part of Knoxville's business community without Scripps Howard. (Courtesy of Bill Hayes.)

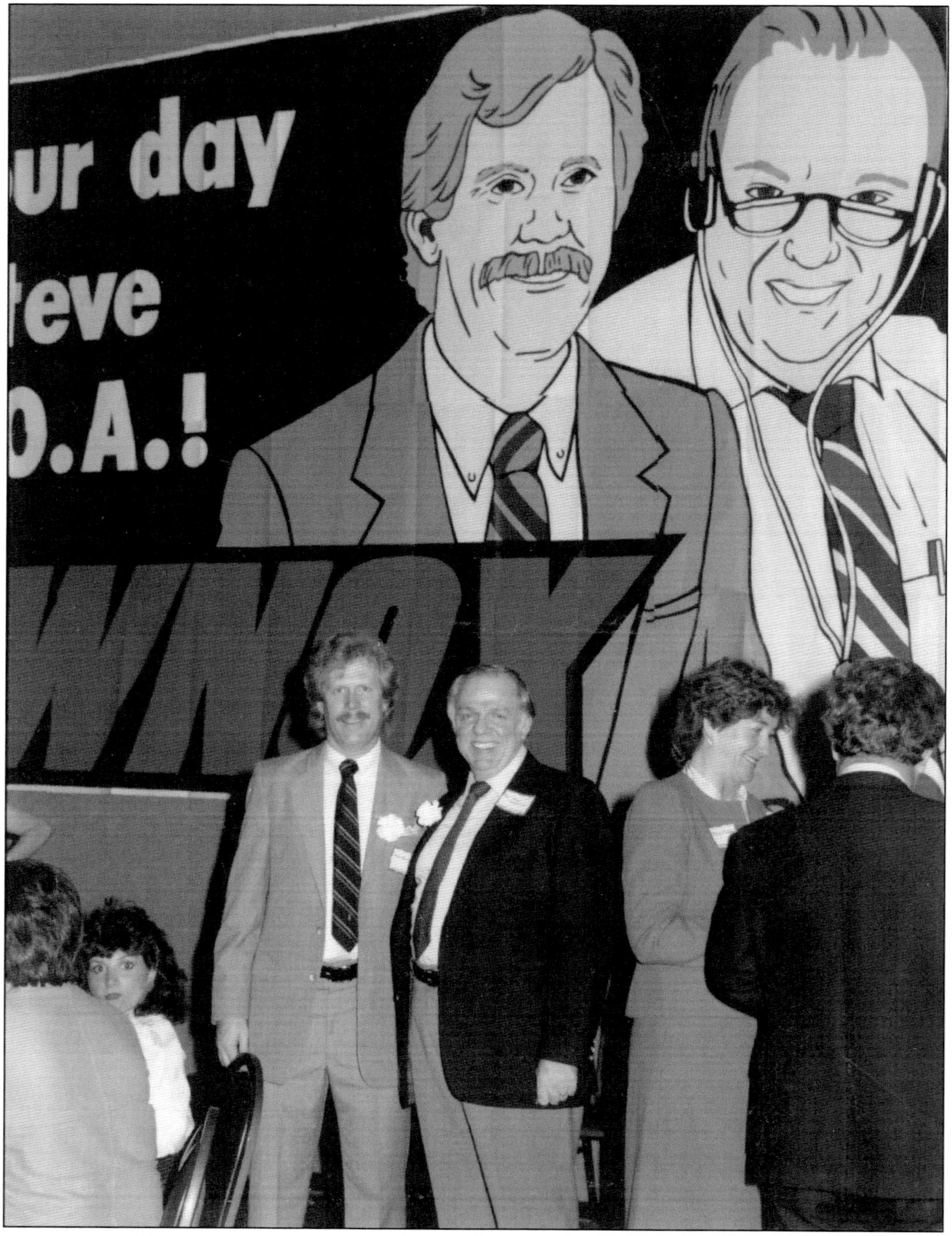

Pictured is WNOX morning announcer Steve Harrison (left) with Paul Oscar Anderson. They are standing in front of a WNOX promotional poster with the slogan "Start your day with Steve and POA." Anderson had made a name for himself in Los Angeles broadcasting before moving to Knoxville. (Courtesy of Bill Hayes.)

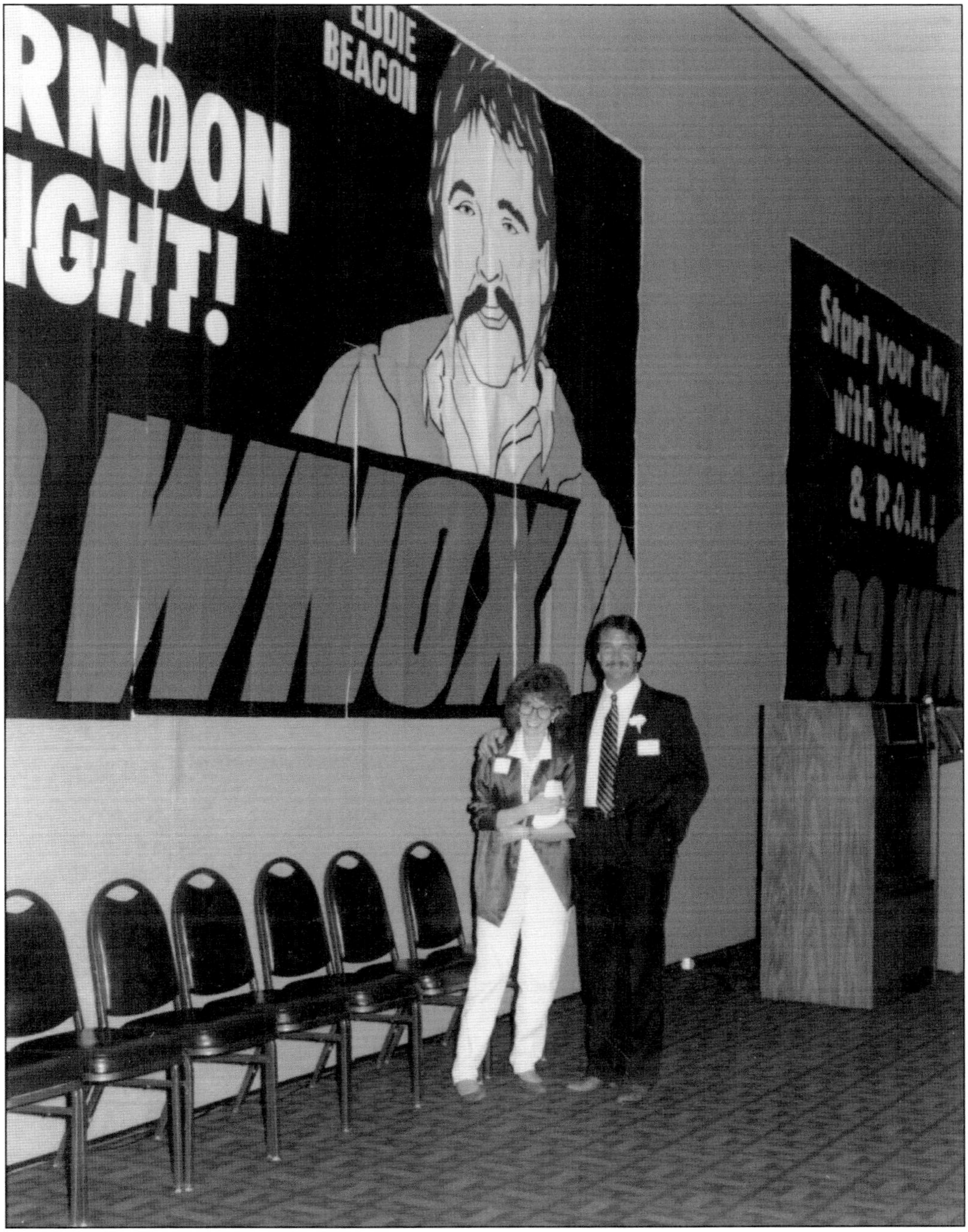

On the right is Eddie Beacon. Although known as "Your Swinging Deacon" in the 1960s as part of the "99ers" who lost his job at the station for an on-air slip by a visiting announcer, he made a comeback at the station as an afternoon announcer. (Courtesy of Bill Hayes.)

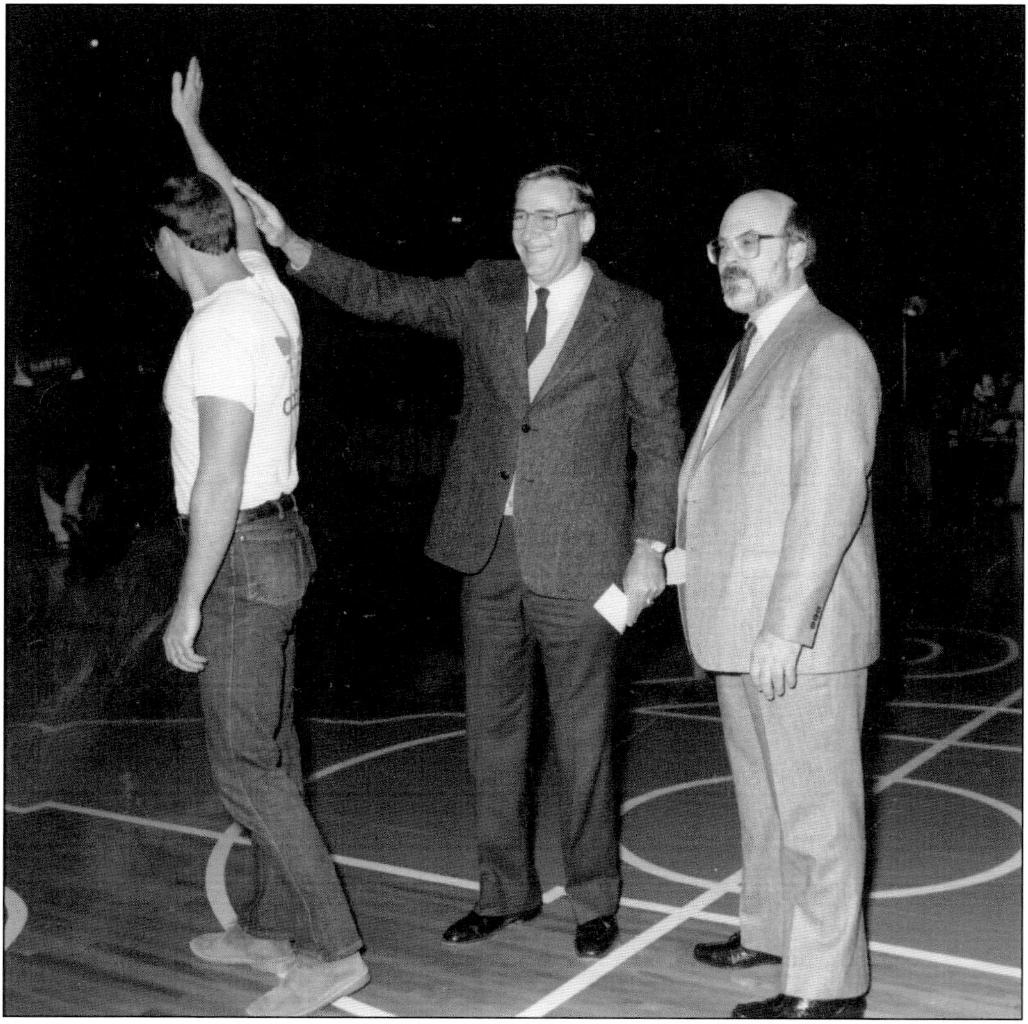

WNOX was still a big University of Tennessee sports station carrying broadcasts of the UT basketball games. This photograph was taken of the winner of a University of Tennessee football field goal kicking contest receiving a check during a UT basketball game. The station did not carry UT football so chose the basketball game to present the check during a half-time ceremony. Second from left is former UT athletic director Doug Dickey, and third from left is WNOX general manager Bill Hayes. (Courtesy of Bill Hayes.)

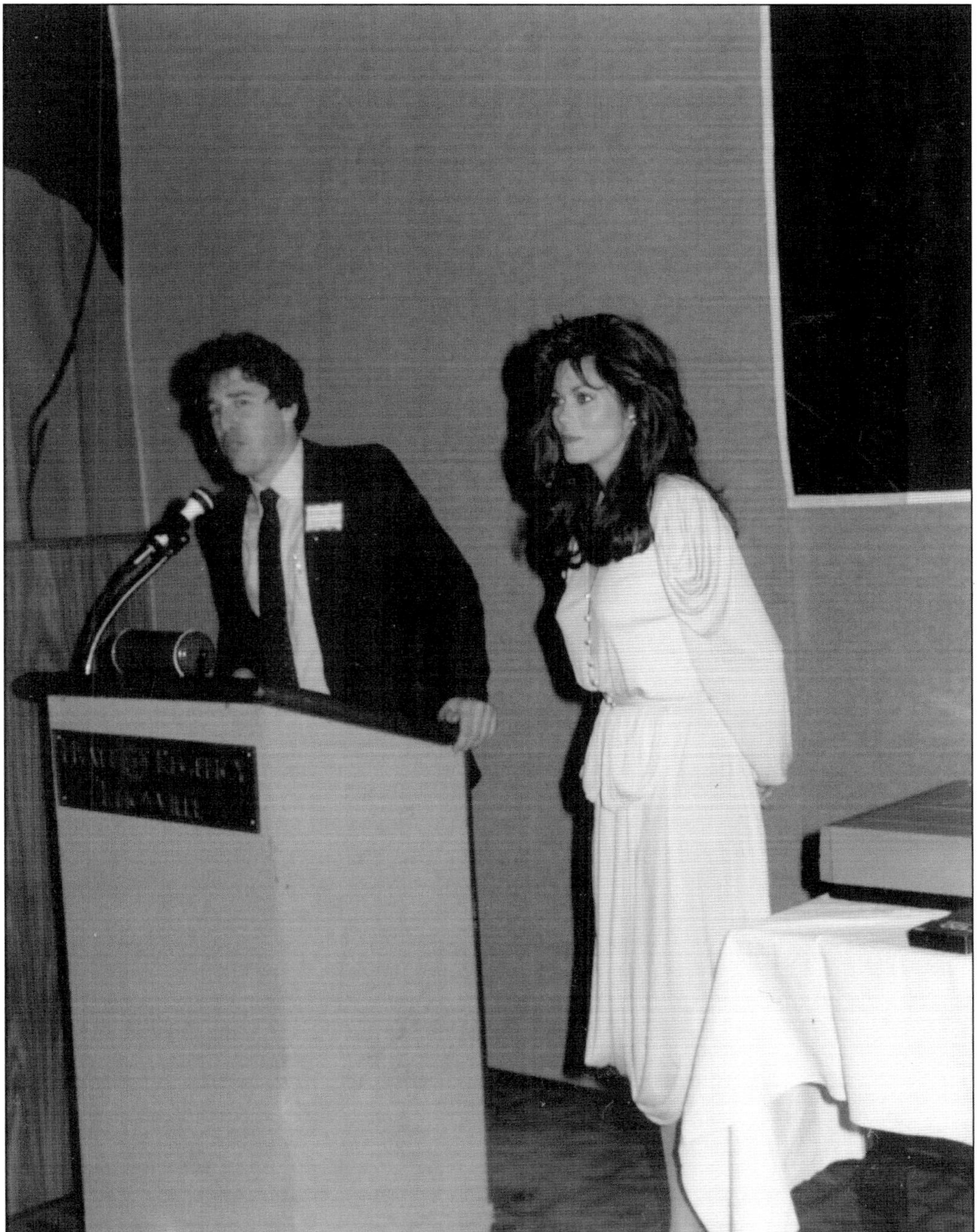

The television program *Dallas* was the number one watched show in the 1980s. During one station event, WNOX brought in *Dallas* actress Deborah Shelton. From 1984 to 1987, she played Mandy Winger on the series—one of lead character J. R. Ewing's more notable mistresses. She is pictured here at the then Hyatt Regency downtown addressing the crowd gathered for the promotion. (Courtesy of Bill Hayes.)

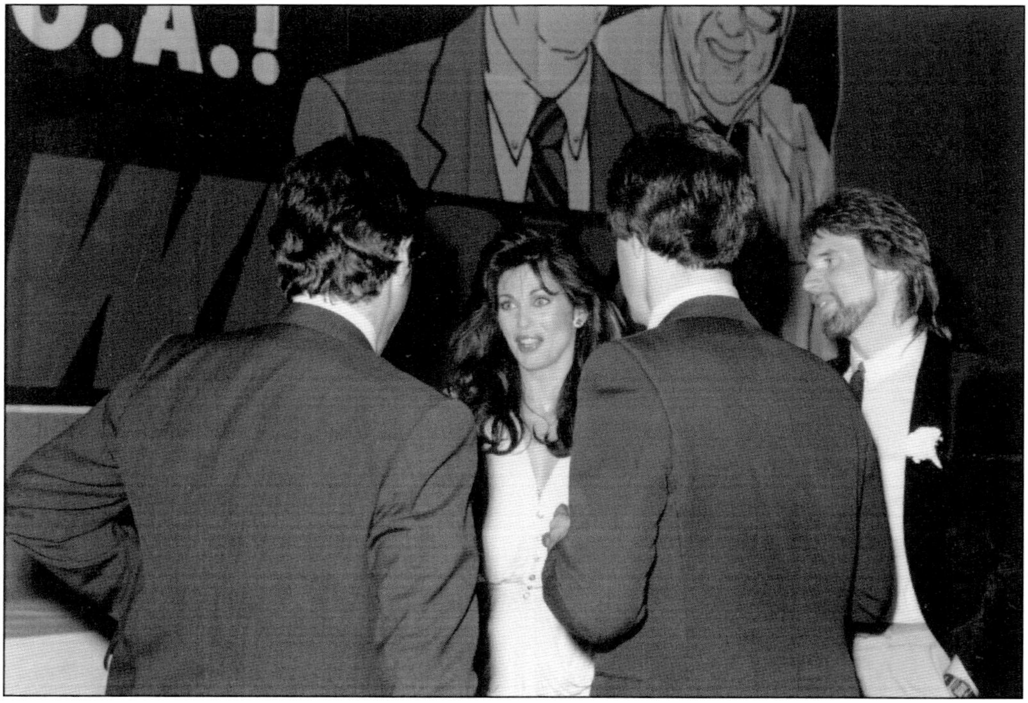

Deborah Shelton is pictured here standing in front of the WNOX banner speaking to fans and signing autographs. While celebrities were in town on station promotions, staff and announcers would have her on the air and the receptions would double as client parties for advertisers. (Courtesy of Bill Hayes.)

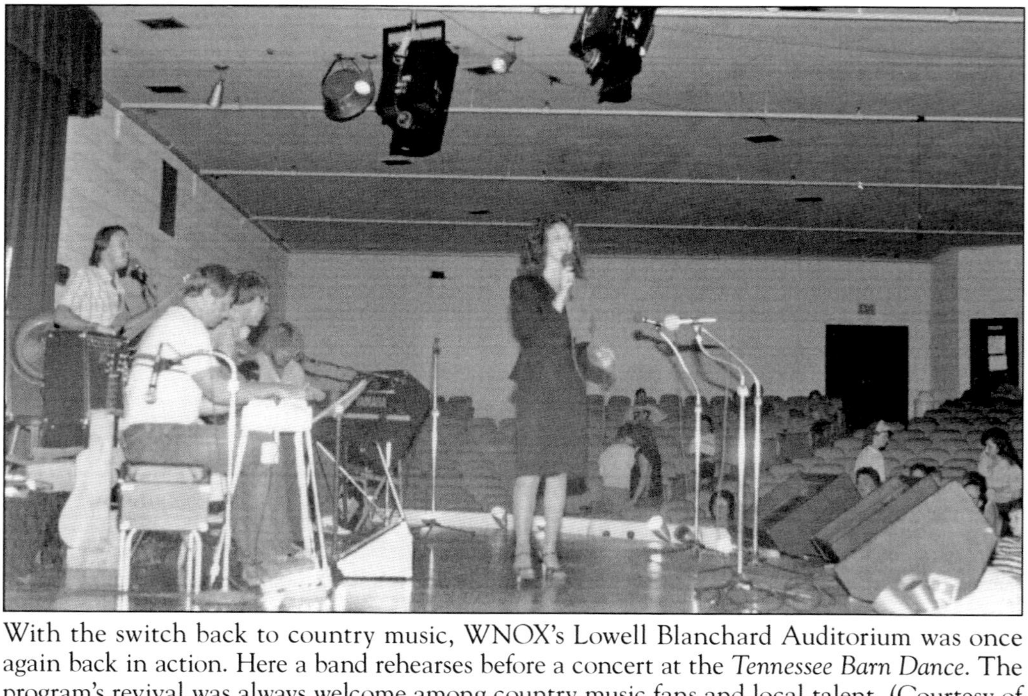

With the switch back to country music, WNOX's Lowell Blanchard Auditorium was once again back in action. Here a band rehearses before a concert at the *Tennessee Barn Dance*. The program's revival was always welcome among country music fans and local talent. (Courtesy of Edith Kirkland.)

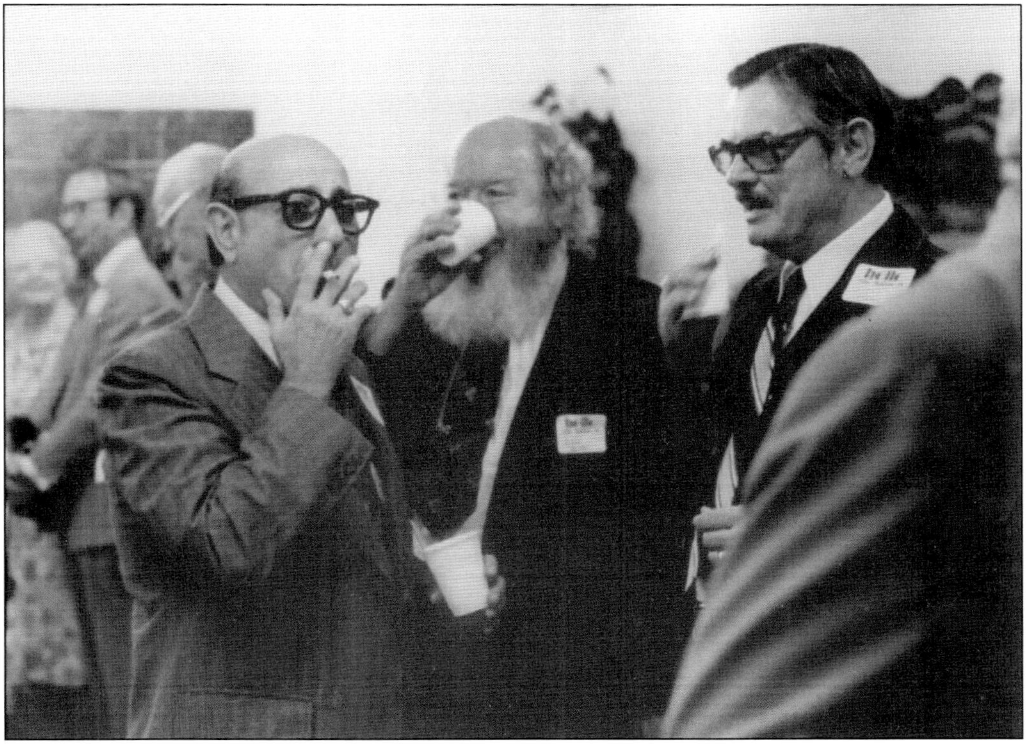

The reunions were still happening in the 1980s for the original *Mid-Day Merry-Go-Round* musicians and staff. In the above photograph, Tony Musco (far left) and Hubert Carter (drinking from the cup) get together in 1983 with old friends to see how they are doing and catch up on the time apart. Most were professional musicians who moved on to other careers. Below, from left to right are musicians Red Kirk, Country Music Association Hall of Famer Bill Carlisle, James Carson, and George Riddle. (Both courtesy of Harley Sweet.)

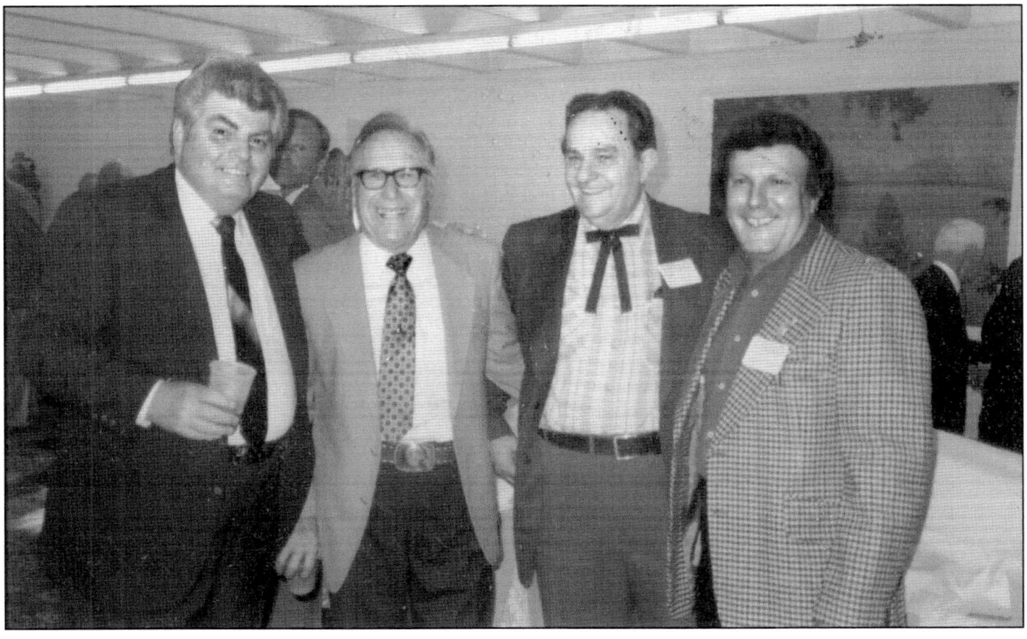

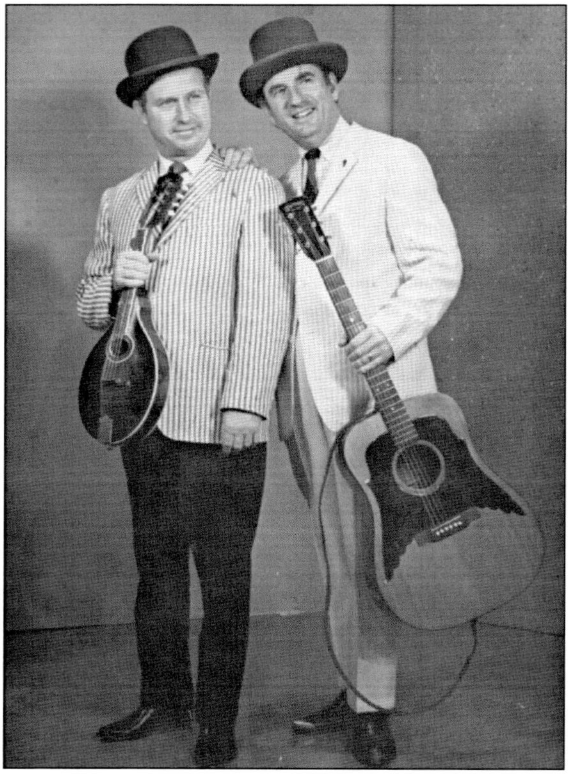

Red Rector (left) and Fred Smith worked as a team for a number of bands on WNOX's *Mid-Day Merry-Go-Round* and *Tennessee Barn Dance*, including Carl Story and the Rambling Mountaineers. Rector went on to become a respected musician and was a fixture in Knoxville's music community until his death in 1990.

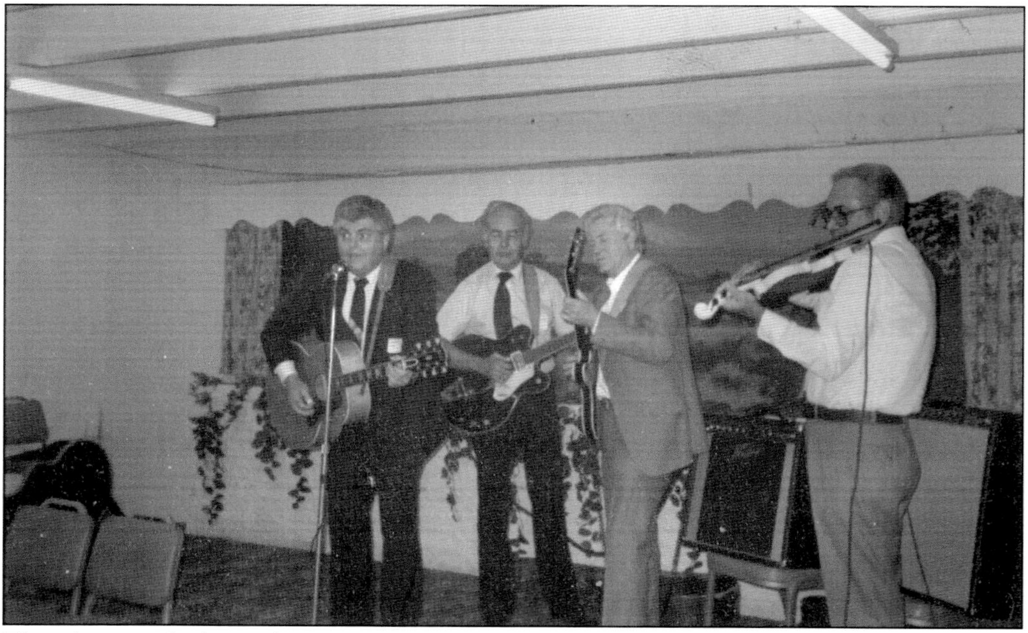

This photograph shows the musicians still knew how to take the stage and perform. From left to right are Red Kirk, Roy Sneed, Luke Brandon, and "Bud" Osborne. The eventual demise of WNOX 990 AM was a striking blow to them. Only through the efforts of promoter John Hitch has a version of the *Tennessee Valley Barn Dance* survived. (Courtesy of Harley Sweet.)

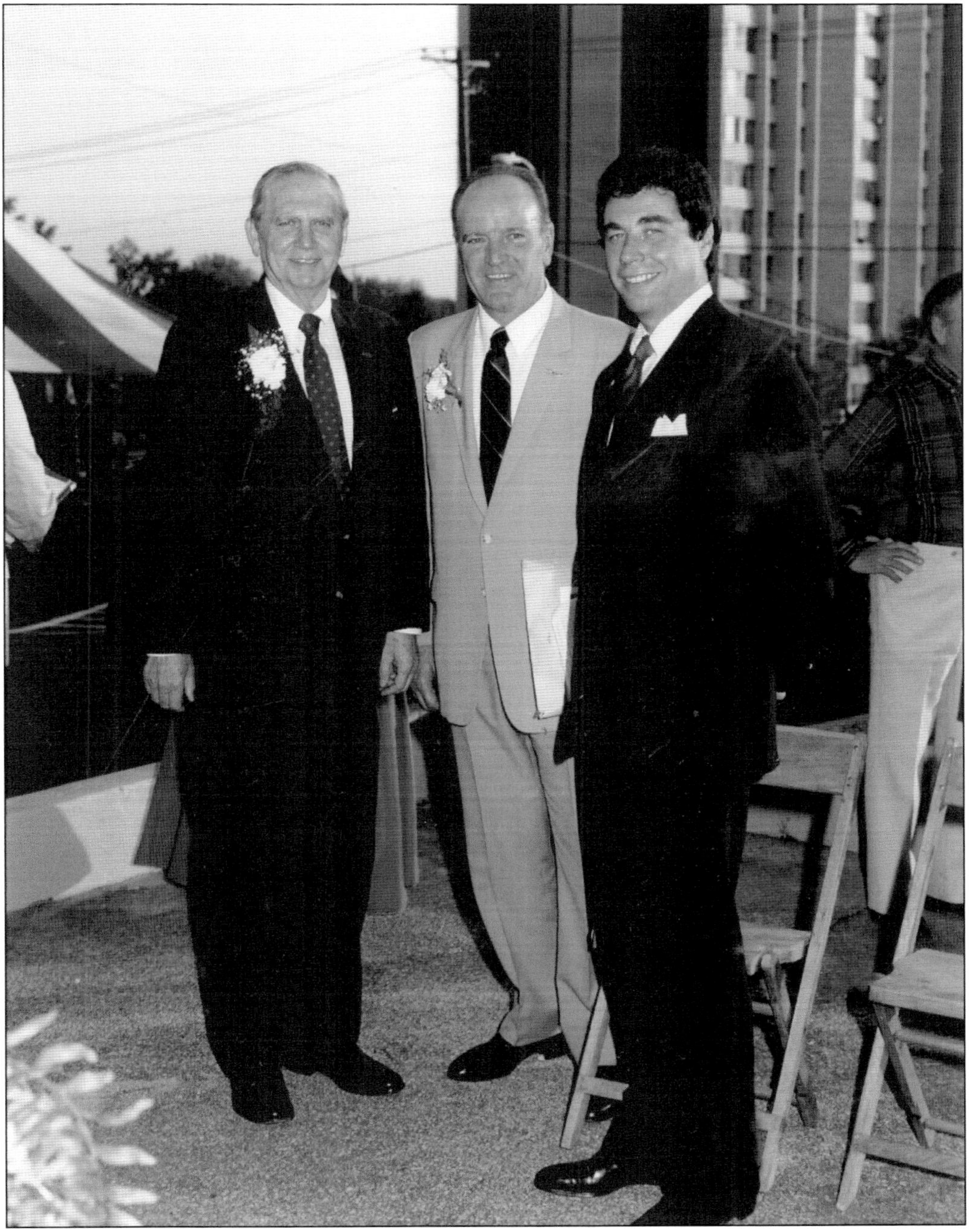

This photograph was taken during the elaborate WNOX Auditorium rededication, where it was named in honor of former WNOX announcer and country music promoter Lowell Blanchard. From left to right are Congressman John Duncan Sr., Mayor Kyle Testerman, and John Lack of ELF Communications. (Courtesy of John Hitch.)

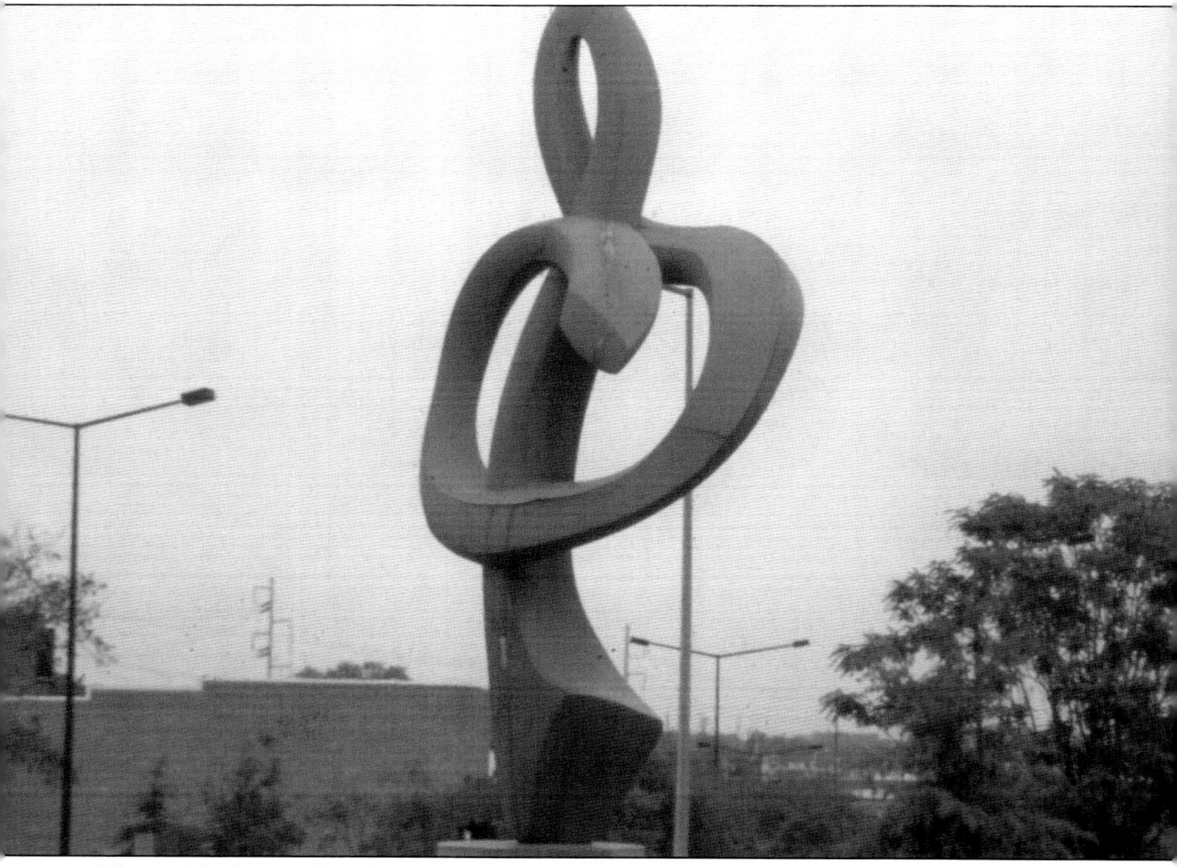

The Treble Clef in Crutch Park is a monument placed by Knox Heritage to commemorate the legacy of Knoxville being the "Cradle of Country Music" and is reminiscent of Lowell Blanchard's comment, "I train them and Nashville gets them." It marks stop 9 of a 19-stop tour through the city. It was dedicated in the 1980s, and many of the plaques lead back to the Gay Street Studios of WNOX AM 990. WNOX was located in one downtown location or another from its founding in 1921 until its departure for the Whittle Springs Studio in 1955, where it still reigned as one of the nation's most notable broadcast organizations.

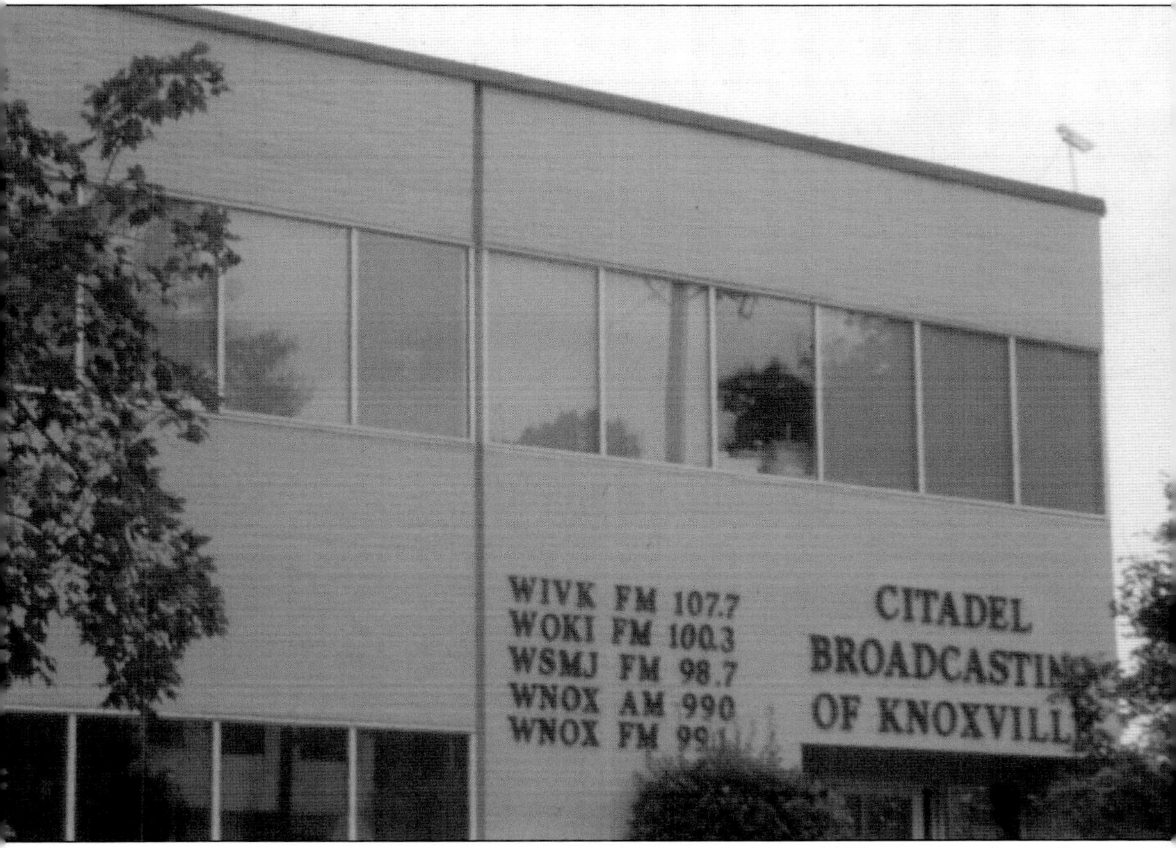

This is the current home of WNOX, now sitting on the 100.3 FM signal. The 990 signal is still in the Citadel Building as an all-sports radio station. The radio station made the switch to news/talk in 1997 and has flourished with the format. Two local announcers, Hallerin Hill and former WNOX employee Phil Williams, bookend the daytime programming of news, information, and weather with local talk shows. The building off of Old Kingston Pike is the headquarters of powerhouse station WIVK.